OXFORD MATHEMATICAL MONOGRAPHS

OXFORD MATHEMATICAL MONOGRAPHS

A. Belleni-Moranti: *Applied semigroups and evolution equations*
M. Rosenblum and J. Rovnyak: *Hardy classes and operator theory*
J. W. P. Hirschfeld: *Finite projective spaces of three dimensions*
A. Pressley and G. Segal: *Loop groups*
J. C. Lennox and S. E. Stonehewer: *Subnormal subgroups of groups*
D. E. Edmonds and W. D. Evans: *Spectral theory and differential operators*
Wang Jianhua: *The theory of games*
S. Omatu and J. H. Seinfeld: *Distributed parameter systems: theory and applications*
J. Hilgert, K. H. Hofmann, and J. D. Lawson: *Lie groups, convex cones, and semigroups*
S. Dineen: *The Schwarz lemma*
B. Dwork: *Generalized hypergeometric functions*
S. K. Donaldson and P. B. Kronheimer: *The geometry of four-manifolds*
T. Petrie and J. Randall: *Connections, definite forms, and four-manifolds*
R. Henstock: *The general theory of integration*
D. W. Robinson: *Elliptic operators and Lie groups*
A. G. Werschulz: *The computational complexity of differential and integral equations*
P. N. Hoffman and J. F. Humphreys: *Projective representations of the symmetric groups*
I. Györi and G. Ladas: *The oscillation theory of delay differential equations*
J. Heinonen, T. Kilpelainen, and O. Martio: *Non-linear potential theory*
B. Amberg, S. Franciosi, and F. de Giovanni: *Products of groups*
M. E. Gurtin: *Thermomechanics of evolving phase boundaries in the plane*
I. Ionescu and M. Sofonea: *Functional and numerical methods in viscoplasticity*
N. Woodhouse: *Geometric quantization* second edition
U. Grenander: *General pattern theory*
J. Faraut and A. Koranyi: *Analysis on symmetric cones*
I. G. Macdonald: *Symmetric functions and Hall polynomials* second edition
B. L. R. Shawyer and B. B. Watson: *Borel's methods of summability*
D. McDuff and D. Salamon: *Introduction to symplectic topology*
M. Holschneider: *Wavelets: an analysis tool*
Jacques Thévenaz: *G-algebras and modular representation theory*
Hans-Joachim Baues: *Homotopy type and homology*
P. D. D'Eath: *Black holes: gravitational interactions*
R. Lowen: *Approach spaces: the missing link in the topology–uniformity–metric triad*

Approach Spaces

The Missing Link in the Topology–Uniformity–Metric Triad

R. LOWEN

Department of Mathematics and Computer Science
University of Antwerp

CLARENDON PRESS • OXFORD
1997

This book has been printed digitally and produced to a standard design in order to ensure its continuing availability

OXFORD
UNIVERSITY PRESS

Great Clarendon Street, Oxford OX2 6DP

Oxford University Press is a department of the University of Oxford.
It furthers the University's objective of excellence in research, scholarship,
and education by publishing worldwide in

Oxford New York

Auckland Bangkok Buenos Aires Cape Town Chennai
Dar es Salaam Delhi Hong Kong Istanbul Karachi Kolkata
Kuala Lumpur Madrid Melbourne Mexico City Mumbai Nairobi
São Paulo Shanghai Singapore Taipei Tokyo Toronto

with an associated company in Berlin

Oxford is a registered trade mark of Oxford University Press
in the UK and in certain other countries

Published in the United States
by Oxford University Press Inc., New York

Reprinted 2002

A catalogue record for this book is available from the British Library

Library of Congress Cataloging in Publication Data
(Data available)

ISBN 0-19-850030-0

To Eva and Wendy

PREFACE

In this book we give a new and complete development of the fundamentals of the theory of approach spaces as originally introduced by the author between 1987 and 1989. Approach spaces form a supercategory of the categories of topological spaces and of metric spaces, and the main purpose for their introduction was to solve the problem of the nonmetrizability of arbitrary initial structures of metrizable topological spaces. Thus, in general, an approach structure is the richest structure, overlying the initial topology, which one can preserve when making such initial structures. Apart from this fundamental aspect, it turns out, moreover, that approach spaces are interesting objects to study in their own right. There are many natural examples of approach structures in various mathematical fields such as probability theory, functional analysis, and theories of function spaces, hyperspaces, and probabilistic metric spaces, and, furthermore, approach spaces seem to be the right setting for a unified treatment of certain topological and metric properties, and for the development of a fundamental theory of approximation.

There are several people to whom I am indebted for a variety of reasons.

It was Horst Herrlich who suggested that I should write a monograph on approach spaces. His continued interest and support have meant a great deal to me. Over the years, the "categorical topology community" in general has been a stimulating environment.

I cooperated with my wife Eva Lowen-Colebunders on the characterization of approach spaces by means of limit operators and the development of several categorically better behaved supercategories of **AP**, and with Piet Wuyts on the description of simultaneously bireflective and bicoreflective subconstructs of **AP**.

My former Ph.D. students Rony Baekeland and Kristin Robeys helped to develop several topics of the theory. My present Ph.D. students Mark Sioen, David Vaughan, Francis Verbeeck, and Bart Windels are working on further developments of either theory or applications of approach spaces. I refer to the bibliography for an account of their contributions to date. They endured the first draft of this monograph and did much of the proofreading. Any errors which may still remain in this work are of course entirely and exclusively my own responsibility.

Finally, I would like to thank the staff at Oxford University Press for their professional help and cooperation in producing this book.

Antwerp R. L.
October 1996

CONTENTS

INTRODUCTION

The ideas developed in this work have their origin in the following considerations.

1 Topological spaces versus metric spaces

There is no question that metric spaces occupy a dominant place in mathematics, not only in the auxiliary form of metrizable topological spaces but also in their own right, as for instance in the theory of Banach spaces. Nevertheless it is topology which provides the framework in which most of the basic concepts of analysis are defined, such as, for example, convergence, continuity, and compactness. One of the reasons why topologies "took over" from metrics is that **TOP** allows for all usual constructions such as subspaces, products, quotients, and coproducts, whereas **MET** (the construct of metric spaces and nonexpansive maps) is stable only under the formation of subspaces and finite products. Although this can easily be remedied by going from metrics to extended pseudometrics ("metrics" where the value infinity is allowed and which do not necessarily have an underlying topology which is Hausdorff), i.e. by going to the construct $p\mathbf{MET}^{\infty}$, a fundamental problem which remains is that metric initial structures do not necessarily accord with topological initial structures and hence are virtually useless. In view of the importance of initial structures, this explains why in many areas of mathematics, out of necessity, one often has to abandon an originally metric setup to get to the setting of topological spaces.

The considerable loss of structure when going from metrics to topologies was partly compensated by the development of uniform spaces by Weil [117] and Dieudonné [28]. Uniformities make it possible to extend some of the concepts available in metric spaces, such as total boundedness and completeness, to the realm of, for instance, locally convex spaces and topological groups. One could argue, however, that when Weil introduced his fundamental concept of uniformities, he took two steps in one, the first being an abstraction and the second a generalization. The association of the set of entourages to a given metric could be interpreted as making an abstraction of the numerical and quantified information of the metric, only to retain uniform information on a set-theoretical level. The description of the axioms of the set of entourages, and the removal of countability conditions, could be interpreted as the generalization part.

The following diagram illustrates the situation.

$$\begin{array}{ccccc}
p\mathbf{MET}^{\infty} & \longrightarrow & ? & \longrightarrow & ? \\
\downarrow & & \downarrow & & \downarrow \\
pm\mathbf{UNIF} & \longrightarrow & \mathbf{UNIF} & \longrightarrow & q\mathbf{UNIF} \\
\downarrow & & \downarrow & & \downarrow \\
pm\mathbf{TOP} & \longrightarrow & \mathbf{CREG} & \longrightarrow & \mathbf{TOP}
\end{array}$$

On the first row we find $p\mathbf{MET}^{\infty}$, the construct of extended pseudometric spaces and nonexpansive maps. On the second row we find $q\mathbf{UNIF}$, the construct of quasi-uniform spaces and uniformly continuous maps, and its full subconstructs **UNIF** of uniform spaces, and $pm\mathbf{UNIF}$ of uniform spaces with a countable basis. On the third row we find **TOP**, the construct of topological spaces and continuous maps, and its full subcategories **CREG** of completely regular spaces, and $pm\mathbf{TOP}$ of pseudometrizable topological spaces.

The first row is the metric level where we find numerical information. The second row is the uniform level where the numerical information has been reduced to uniform information in terms of collections of sets (e.g. the collection of entourages). The third row is the topological level where the uniform information has been further reduced to local information, also in terms of collections of sets (e.g. the collections of neighborhoods of points). The four meaningful vertical arrows stand for the forgetful functors and the four meaningful horizontal arrows stand for the embedding functors.

As we can see, however, the diagram is incomplete; why is this so? Does there exist a type of structure which would allow us to fill in the two question marks in such a way that the two new vertical arrows also represent forgetful functors, that the two new horizontal arrows also represent embedding functors, and that the whole diagram commutes?

2 Products of metric spaces: the heart of the matter

There is a natural forgetful functor

$$p\mathbf{MET}^{\infty} \longrightarrow \mathbf{TOP}$$

which takes an ∞p-metric space to its underlying topological space and leaves morphisms unaltered. This underlying topology functor commutes with the formation of finite products. If we take a finite family of metric spaces and make their product in $p\mathbf{MET}^{\infty}$ then the underlying topology is the product topology. Loosely speaking, a finite product of metrizable topological spaces is "canonically metrizable" once the metrics on the factors are fixed. The underlying topology functor, however, does not commute with infinite products. There are two cases to be considered: countable products and uncountable products. If we consider a countable family of ∞p-metric spaces $(X_n, d_n)_{n\in\mathbb{N}}$ then the product topology on $\prod_{n\in\mathbb{N}} X_n$ can be metrized by a host of *ad hoc* ∞p-metrics of various type as, for example,

$$d(x,y) \doteq \sum_{n\in\mathbb{N}} a_n(d_n(x_n,y_n) \wedge b)$$

or

$$d(x,y) \doteq \sup\{a_n(d_n(x_n,y_n) \wedge b) \mid n \in \mathbb{N}\},$$

for suitable choices of sequences $(a_n)_n$ and $(b_n)_n$, but it is never metrized by the product ∞p-metric in $p\mathbf{MET}^{\infty}$. As soon as there are more than a countable number of factors, the product space simply is no longer metrizable, neither by the product ∞p-metric in $p\mathbf{MET}^{\infty}$, nor by any other ∞p-metric.

Why can we produce a canonical ∞p-metric for finite products, only ad hoc ∞p-metrics for countable products, and no ∞p-metric for uncountable products? Is it really the case that unless we have a finite number of factors all canonical notion of distance in the product space has simply vanished?

3 Analysis versus numerical analysis and approximation theory

Consider the real line $\mathbb{R}$ with the usual Euclidean metric and topology, and consider the sequences $(x_n)_n$ and $(y_n)_n$ where

$$x_n \doteq \begin{cases} \epsilon & n \text{ even,} \\ -\epsilon & n \text{ odd,} \end{cases} \quad \text{and} \quad y_n \doteq \begin{cases} n & n \text{ even,} \\ -n & n \text{ odd.} \end{cases}$$

Neither of these sequences converges. The first one, however, has two main convergent subsequences, and from the point of view of numerical analysis or approximation theory, for a "sufficiently small" ϵ, the sequence itself might actually be considered "sufficiently" convergent, e.g. to 0. The second sequence on the other hand has no convergent subsequences, and could not even remotely be considered to be "approximately convergent" to any point of $\mathbb{R}$.

A more striking example is obtained as follows. Let $\varphi : \mathbb{R} \longrightarrow]-\epsilon, \epsilon[$ be a homeomorphism, and let $(r_n)_n$ be an enumeration of the rationals. The sequence $(r_n)_n$ is not remotely "approximately convergent" to any point of $\mathbb{R}$. The sequence $(\varphi(r_n))_n$ on the other hand, for a "sufficiently small" ϵ, might again be considered "sufficiently" convergent.

By means of the topology of $\mathbb{R}$, not only can we not detect the different behaviour of these sequences, we must conclude that they are "identical". In order to "see" the difference we require the metric. In the literature on approximation theory concepts have been introduced to deal with these ideas in the setting of metric (mainly normed) spaces, albeit in an *ad hoc* way. We refer to the work of Edelstein [33], Anderson *et al.* [2], Lim [65], and Holmes [52]. There is, however, no canonical theory by means of which we can classify sequences according to the above criteria, in the way topological convergence classifies sequences in convergent and nonconvergent sequences. We also refer to the paper by Linz [66] for a fundamental critique on this matter in the context of numerical analysis. More fundamentally, however, once we leave the realm of metric spaces we have no mathematical model, canonical or *ad hoc*, to cope with these ideas.

Is it possible to build a theory in which the above ideas in the realm of approximation theory arise as naturally as the theory of convergence arises in the setting of topology?

4 Topological properties versus metric properties

Certain topological properties have remarkable metric counterparts. Compactness is a topological property which in the case of a metrizable topological space is closely linked to the metric concept of total boundedness. But more strikingly, both properties have similar characterizations. For compactness one starts with arbitrary open covers and for total boundedness one starts with collections of all balls with an arbitrary fixed radius, and in both cases, for a space to be respectively compact or totally bounded, one asserts the existence of a finite subcollection which still covers the whole space.

A similar situation presents itself for connectedness. Topological connectedness of a space X can be characterized by the fact that X cannot be split into two nontrivial closed parts. Similarly, Cantor-connectedness in metric spaces (introduced by Cantor in [23] and extended to the realm of uniform spaces by Mrowka and Pervin in [89]) is characterized by the fact that the space cannot be split into two nontrivial parts which lie at a strictly positive distance from each other.

Is there a unifying theory behind these different but similar concepts?

5 Quantification of topological properties

In [62] Kuratowski introduced a measure of noncompactness for subsets of a complete metric space. Since then several variants have been introduced in the literature. Probably the nicest and most widely used measure is the so-called ball measure or Hausdorff measure of noncompactness, see e.g. Nussbaum [96]. For an extensive account on applications of these measures in Banach spaces we refer to the book by Banaś and Goebel [8]. Remarkably, for all these measures, the deviation a subset has from being compact is actually more the deviation it has from being totally bounded. So here too one sees a certain blending of topological ideas with metric concepts.

Is there a canonical theory where quantified topological concepts exist in the same way as topological concepts exist in **TOP***?*

It is our purpose in this work to show that all the points raised in the five items above can be addressed in a satisfactory way and that all the problems mentioned can be solved within the setting of what we call approach spaces.

In the first chapter we introduce the different structures by which we can define the construct **AP** of approach spaces. This should be compared to the first chapter of a book on topology where the different axiom systems of open sets, closed sets, closure operator, interior operator, neighborhood system, and convergence are defined and are proved to be equivalent. Here too we have to prove that the different structures by means of which we can define an approach space (seven in total) are equivalent. As is the case in topology, some transitions from one structure to another are both pleasing and fundamental; others may be somewhat circuitous and are of lesser importance. Further, in this first chapter we also define the morphisms in **AP** and we show that **AP** is a nice construct; more precisely we show that it is a topological construct in the sense of Adámek *et al.* [1] (see also Appendix A).

In the second chapter we prove that **TOP** can be viewed as a nice subconstruct of **AP**. Precisely we show that it can be embedded as a simultaneously bireflective and bicoreflective concrete subconstruct. The fact that **TOP** is bicoreflectively embedded is especially important since it will allow us to interpret the **TOP**-coreflection of an approach space in the same way that we interpret the topology underlying a metric. Moreover, in this chapter we also completely characterize all other simultaneously bireflective and bicoreflective concrete subconstruct of **AP**.

In the third chapter we prove that $p\mathbf{MET}^{\infty}$ also can be viewed as a nice subconstruct of **AP**. Precisely we show that it can be embedded as a bicoreflective concrete subconstruct. Here it is the fact that $p\mathbf{MET}^{\infty}$ is *not* reflectively embedded in **AP**; more precisely that it is not stable under the formation of arbitrary initial structures, which is important, since this allows us, in the fourth chapter, to solve the problems raised in the first and second items. In the third chapter we also tackle the points raised in the third item concerning the existence of a canonical theory of approximate convergence.

In the fourth chapter we characterize a subconstruct of **AP**, namely the epireflective hull of $p\mathbf{MET}^{\infty}$, consisting of what we call uniform approach spaces and which we denote by **UAP**. This allows us to solve completely the problems raised in the first and second items and also allows us to fill in the spaces of the question marks in the diagram of the first item.

In the fifth chapter we show that, although, at least until recently, approach spaces did not explicitly exist in the literature, they nevertheless existed implicitly, in the sense that many well-known topologies and uniformities are actually defined by means of an approach structure. We look at canonical and uniquely defined approach structures which induce the weak topology on spaces of probability measures (see e.g. Billingsley [13]), the topologies of $\mathfrak{S}$-convergence (in the sense of Bourbaki [18]) on function spaces, the Wijsman topology on hyperspaces of metric spaces (see e.g. Lechicki and Levi [64]), the strong topology on probabilistic metric spaces (see e.g. Schweizer and Sklar [106]), and the topology of convergence in probability on random variables (see e.g. Billingsley [13] and Sherwood [110]).

In the sixth chapter we address the points raised in the fourth and fifth items. We show that compactness and total boundedness are special instances of a unifying property in **AP** and we show that connectedness and Cantor-connectedness too are special instances of a unifying property in **AP**. Moreover, we show that the Hausdorff measure of noncompactness generalizes to a canonical measure of compactness in **AP**. Following the same philosophy we construct a measure of connectedness in **AP** which has equally good properties as the measure of compactness. In this chapter we also introduce a notion of completeness in approach spaces, which will turn out to coincide with usual completeness in the case of metric spaces and which gives some surprising new views on completeness of, for example, function spaces and hyperspaces.

Since **AP** contains both topological spaces and metric spaces we are in the peculiar situation where for uniform approach spaces we can construct both a completion and a compactification. In the seventh and eighth chapters we develop both these extensions. They will both turn out to be categorically right in the sense that they are epireflections on suitable subconstructs of respectively complete and compact objects. It will be shown that they are the right counterparts for the completion of a metric space in $p\mathbf{MET}^{\infty}$ and for the Čech–Stone compactification of Tychonoff spaces in **TOP**. In particular in the last chapter we will see that our compactification allows us to equip $\beta\mathbb{N}$ with a canonical approach structure which generates the Čech–Stone topology and which extends the usual metric on $\mathbb{N}$, an impossible situation in the usual topological setting.

In the literature one can find many generalizations of the notion of a metrizable topological space and equally many generalizations of the notion of a metric space. A not so recent, but nevertheless still very interesting, survey is given by Burke and Lutzer in [22]. More recent survey papers are those by Gruenhage [43], [44] and Nagata [91].

In the second group, mainstream research has dealt with replacing the real numbers as codomain for a generalized metric by more general spaces, while retaining the premise that a generalized metric gives a distance between pairs of points. For fundamental work in this area we refer to the papers of Reichel and Ruppert [103] and Reichel [102].

In the first group, several notions generalizing metrizability and involving some kind of "distance functions" have been introduced over the years. The one which is most closely related to our concept of approach structure is the concept of a κ-metric as introduced by Shchepin [107]. This concept was further studied by Shchepin himself in [108], [109], by Isiwata in [56], [57], [58], and by Suzuki *et al.* in [113] among others. A Tychonoff space X is called κ-metrizable if there exists a function $\rho : X \times RC(X) \longrightarrow \mathbb{R}$, where $RC(X)$ stands for the collection of regularly closed sets in X, fulfilling the properties:

1. $\rho(x, F) = 0$ if and only if $x \in F$,
2. $F \subset F' \;\Rightarrow\; \rho(x, F') \leq \rho(x, F), \;\; \forall x \in X,$

3. $\rho(., F)$ is continuous, $\forall F \in RC(X)$
4. $\rho(x, \mathrm{cl}(\mathrm{int}(\bigcup_{j \in J} F_j))) = \inf_{j \in J} \rho(x, F_j)$, $\forall (F_j)_j$ increasing totally ordered.

Such a function is called a κ-metric for X. The class of κ-metrizable topological spaces contains all metrizable topological spaces and is closed under the formation of products. This work was the first to address the problem of the nonmetrizability of arbitrary products of metrizable spaces and to give a partial solution to it. However, it did not solve the problem as formulated under the first and second items, i.e. in the categorical and canonical sense, and it also did not address the third, fourth, and fifth items. Analogous ideas also appeared in the work of Borges [15], Nagata [91], and Naimpally and Pareek [92] where the concept of an annihilator was used. An annihilator basically is more general than a κ-metric, only fulfills the first and third properties above, and can be defined on $X \times C(X)$ where $C(X)$ is an arbitrary collection of closed sets. Much work in this area was aimed at finding supplementary conditions on a κ-metric or an annihilator to insure metrizability of the given space, see e.g. the paper of Suzuki *et al.* [113].

1
APPROACH SPACES

In this chapter we introduce the basic structures which determine what we will call an approach space. One of the powerful features of approach spaces is that they can be determined by conceptually very different but nevertheless equivalent structures. This is not unlike the situation in topology, where the structure of a topological space can be determined by a number of equivalent concepts, such as, for example, open sets, closed sets, closure operator, neighborhood system, and convergence structure. In the next two chapters where we discuss the relation between approach spaces, topological spaces, and metric spaces, the parallelism will become quite apparent.

1.1 DISTANCES

Throughout this work, given a set X, we denote the set of all subsets of X by 2^X. Further we put $\mathbb{R}^+ \doteqdot [0, \infty[$ and $\mathbb{R}_0^+ \doteqdot]0, \infty[$.

Intuitively, probably the most appealing structure which we will be considering is that of a distance between points and sets. Whereas in a metric space (X, d) a distance between pairs of points is given, and a distance between points and sets can then be derived from this according to the usual formula

$$\delta_d(x, A) \doteqdot \inf_{a \in A} d(x, a) \qquad \forall x \in X, \ \forall A \in 2^X,$$

here we consider the latter as a primitive structure, i.e. we start from a concept of distance between points and sets. Before giving the precise definition we need to introduce the following notation.

If X is a set, and we have a function $\delta : X \times 2^X \longrightarrow [0, \infty]$, then for any subset $A \subset X$ and any $\epsilon \in [0, \infty]$, we define

$$A^{(\epsilon)} \doteqdot \{x \in X \mid \delta(x, A) \leq \epsilon\}.$$

1.1.1 Definition A function

$$\delta : X \times 2^X \longrightarrow [0, \infty]$$

is called a *distance (on X)* if it satisfies the following properties:

(D1) $\forall x \in X: \quad \delta(x, \{x\}) = 0,$

(D2) $\forall x \in X: \quad \delta(x, \emptyset) = \infty,$

(D3) $\forall x \in X, \ \forall A, B \in 2^X: \ \delta(x, A \cup B) = \min(\delta(x, A), \delta(x, B)),$

(D4) $\forall x \in X, \ \forall A \in 2^X, \ \forall \epsilon \in [0, \infty]: \ \delta(x, A) \leq \delta(x, A^{(\epsilon)}) + \epsilon.$

In the same way as in a metric space the value $\delta(x, A)$ is interpreted as the distance from the point x to the set A. We wish to emphasize, however, that this distance is not necessarily derivable from the distances $\delta(x, \{a\})$ for $a \in A$. We will give numerous examples of such instances in the rest of this work.

The codomain of a distance and also of many other functions which we will consider in this work is $[0, \infty]$, and it should be pointed out that we will freely use the symbols $+$ and $-$ for the natural "extensions" of these operations. More precisely, $+$ and $-$ stand for the usual addition and subtraction in the case of real numbers and further, for any $x \in [0, \infty[$ we have $x + \infty = \infty + x = \infty + \infty = \infty$, $\infty - x = \infty$, and $\infty - \infty = 0$.

The following proposition contains some simple but fundamental properties which we will use implicitly in the sequel.

1.1.2 Proposition *If $\delta : X \times 2^X \longrightarrow [0, \infty]$ is a distance then the following hold:*

1. $\forall x \in X, \ \forall A \in 2^X: \ x \in A \Rightarrow \delta(x, A) = 0,$
2. $\forall x \in X, \ \forall A, B \in 2^X: \ A \subset B \Rightarrow \delta(x, B) \leq \delta(x, A),$
3. $\forall x \in X, \ \forall \mathcal{A} \subset 2^X, \ \mathcal{A}$ *finite*: $\ \delta(x, \bigcup \mathcal{A}) = \min_{A \in \mathcal{A}} \delta(x, A),$
4. $\forall x \in X, \ \forall A, B \in 2^X: \ \delta(x, A) \leq \delta(x, B) + \sup_{b \in B} \delta(b, A).$

Proof The second and third properties follow immediately from (D3) and the first property follows from the second property and (D1). To prove the fourth property let $x \in X, \ A, B \in 2^X$, and let

$$\epsilon \doteqdot \inf \left\{ \theta \in [0, \infty] \mid B \subset A^{(\theta)} \right\};$$

then it follows that

$$\begin{aligned}\delta(x, A) &\leq \delta\left(x, A^{(\epsilon)}\right) + \epsilon \\ &\leq \delta(x, B) + \sup_{b \in B} \delta(b, A).\end{aligned}$$

□

Property 4 in the foregoing proposition, modulo the difference in domains for κ-metrics and for distances, is the same as the so-called regularity condition for κ-metrics in Shchepin [109].

Although a distance is defined as a function of two variables, points, and subsets, it will sometimes be useful also to consider the following associated functions on points. For a given subset $A \subset X$ we define

$$\delta_A : X \longrightarrow [0, \infty] : x \longrightarrow \delta(x, A).$$

As is usually done in the context of hyperspaces of metric spaces (see e.g. Lechicki and Levi [64]) we will also call such functions *distance functionals*.

1.2 LIMIT OPERATORS

The next structure which we introduce provides us with a notion of convergence. The basic set-theoretical tools needed to define convergence are sequences, nets, and filters. Sequences suffice in metric spaces, but it will soon be clear that these do not suffice in our case. Although in particular cases we will sometimes give results for sequences, for the general theory we have to resort to either filters or nets. We have chosen filters, as introduced by Choquet in [25], because of their conceptual beauty. All results can, however, be properly rephrased in terms of nets.

Whenever convenient we will use the following notations. $\mathbf{F}(X)$ will stand for the set of all filters on X, and $\mathbf{U}(X)$ will stand for the set of all ultrafilters on X. Sometimes it will be useful to generalize this notation in the following way. If $\mathcal{F}$ is a given filter on X, then we will denote by $\mathbf{F}(\mathcal{F})$ the collection of all filters on X which are finer than $\mathcal{F}$, and by $\mathbf{U}(\mathcal{F})$ the collection of all ultrafilters on X which are finer than $\mathcal{F}$. If $\mathcal{F}$ is the trivial filter on X, i.e. $\mathcal{F} = \{X\}$, then $\mathbf{F}(\mathcal{F})$ reduces to $\mathbf{F}(X)$ and $\mathbf{U}(\mathcal{F})$ reduces to $\mathbf{U}(X)$.

If $\mathcal{A}$ is a collection of subsets of X, then the stack of $\mathcal{A}$ is defined as

$$\mathrm{stack}\mathcal{A} \doteqdot \{B \subset X \mid \exists A \in \mathcal{A}\colon\ A \subset B\}.$$

If $\mathcal{F}$ is a filter on X then the sec of $\mathcal{F}$ is defined as

$$\mathrm{sec}\mathcal{F} \doteqdot \bigcup_{\mathcal{U} \in \mathbf{U}(\mathcal{F})} \mathcal{U} = \{A \subset X \mid \forall F \in \mathcal{F}:\ A \cap F \neq \emptyset\}.$$

If $\mathcal{A}$ is a filterbasis then stack$\mathcal{A}$ is the filter generated by $\mathcal{A}$. In case $\mathcal{A}$ reduces to a single set A we write stack$A$ instead of stack$\{A\}$, and in case the single set A furthermore reduces to a single point a we write stack$a$ instead of stack$\{a\}$. If $A \in 2^X$, $A \neq \emptyset$, then stack$A$ is the filter consisting of all sets containing A, and then, for simplicity, especially in formulas, we will often write $\mathbf{F}(A)$ instead of $\mathbf{F}(\mathrm{stack}A)$, and $\mathbf{U}(A)$ instead of $\mathbf{U}(\mathrm{stack}A)$.

The axioms for a limit operator which we are about to introduce are similar to the axioms for a topological convergence structure, Kent [60], Kowalsky [61], Lowen-Colebunders [84]. One of the axioms requires the so-called *Kowalsky diagonal operation*, Kowalsky [61]. Given any index set J, a collection of filters $(\mathcal{F}_j)_{j\in J}$ on X, and a filter $\mathcal{F}$ on J, *the diagonal filter of the family* $(\mathcal{F}_j)_{j\in J}$ *with respect to* $\mathcal{F}$ is defined as

$$\mathcal{D}\Big((\mathcal{F}_j)_{j\in J}, \mathcal{F}\Big) \doteqdot \bigvee_{F\in\mathcal{F}} \bigcap_{j\in F} \mathcal{F}_j.$$

The symbol $\bigvee$ in this notation stands for the supremum of a collection of filters. Note that in the particular case of a diagonal filter this supremum always exists.

In the special case where the collection of filters is indexed by the points of X, we call this a *selection of filters on* X, and we denote this selection in a functional way by

$$\mathcal{S} : X \longrightarrow \mathbf{F}(X) : x \longrightarrow \mathcal{S}(x).$$

The diagonal filter of $(\mathcal{S}(x))_{x\in X}$ with respect to a filter $\mathcal{F}$ on X is then denoted by $\mathcal{D}(\mathcal{S}, \mathcal{F})$ for short, i.e.

$$\mathcal{D}(\mathcal{S}, \mathcal{F}) = \mathcal{D}((\mathcal{S}(x))_{x\in X}, \mathcal{F}) = \bigvee_{F\in\mathcal{F}} \bigcap_{x\in F} \mathcal{S}(x).$$

1.2.1 Definition A function

$$\lambda : \mathbf{F}(X) \longrightarrow [0, \infty]^X$$

is called a *limit operator (on X)* if it satisfies the following properties:

(L1) $\forall x \in X : \ \lambda(\text{stack}x)(x) = 0$,

(L2) $\forall \mathcal{F}, \mathcal{G} \in \mathbf{F}(X) : \ \mathcal{F} \subset \mathcal{G} \Rightarrow \lambda\mathcal{G} \leq \lambda\mathcal{F}$,

(L3) for any family $(\mathcal{F}_j)_{j \in J}$ of filters on X,

$$\lambda\left(\bigcap_{j\in J} \mathcal{F}_j\right) = \sup_{j\in J} \lambda\mathcal{F}_j,$$

(L4) for any $\mathcal{F} \in \mathbf{F}(X)$ and any selection of filters $(\mathcal{S}(x))_{x\in X}$ on X,

$$\lambda(\mathcal{D}(\mathcal{S}, \mathcal{F})) \leq \lambda\mathcal{F} + \sup_{x\in X} \lambda(\mathcal{S}(x))(x).$$

For any $\mathcal{F} \in \mathbf{F}(X)$, the function

$$\lambda\mathcal{F} : X \longrightarrow [0, \infty]$$

is called the *limit of* $\mathcal{F}$.

The value $\lambda\mathcal{F}(x)$ is interpreted as the distance that the point x is away from being a limit point of the filter $\mathcal{F}$. The word distance here is meant intuitively and not in the sense of 1.1.1. However, apart from the general formulas which exist involving distances and limits and which we will prove in this chapter, in the third and fourth chapters some more specific relationships will be shown to hold between these two notions.

The smaller the value of $\lambda\mathcal{F}(x)$, the closer x comes to being a limit point of $\mathcal{F}$. Of course, at this point this is merely an intuitive and as yet unsupported interpretation of an abstract concept, but both theory and examples will justify the use of this interpretation. Notice also that (L2) actually is a consequence of (L3).

Condition (L4) has recently been generalized in various interesting ways by Brock and Kent [19], [20] in order to investigate a notion of regularity for approach spaces. One of the consequences of their investigations is that they were able to combine (L3) and (L4) into a single condition.

Before moving on to the next structure we give some preliminary properties related to filters. The following is a useful, purely filter-theoretic result which we will require on several occasions.

1.2.2 Proposition *If $\mathcal{F} \in \mathbf{F}(X)$, and for each ultrafilter $\mathcal{U} \in \mathbf{U}(\mathcal{F})$, $\sigma(\mathcal{U}) \in \mathcal{U}$, then there exists a finite set $\mathbf{U}_\sigma \subset \mathbf{U}(\mathcal{F})$ such that*

$$\bigcup_{\mathcal{U} \in \mathbf{U}_\sigma} \sigma(\mathcal{U}) \in \mathcal{F}.$$

Proof Suppose the conclusion does not hold. Then this implies that the family

$$\mathcal{F} \cup \{X \setminus \sigma(\mathcal{U}) \mid \mathcal{U} \in \mathbf{U}(\mathcal{F})\}$$

has the finite intersection property, and thus is contained in some ultrafilter $\mathcal{U} \in \mathbf{U}(\mathcal{F})$. This, however, implies that both $\sigma(\mathcal{U}) \in \mathcal{U}$ and $X \setminus \sigma(\mathcal{U}) \in \mathcal{U}$ which is a contradiction.

□

We will also need the following results concerning the diagonal operation. For more information on the interpretation and use of this operation we refer to Kowalsky [61] and Brock and Kent [19], [20].

1.2.3 Proposition *Let $(\mathcal{F}_j)_{j\in J}$ be a family of filters on X, and let $\mathcal{F}$ be a filter on J; then the following properties hold:*

1. $\mathcal{D}\Big((\mathcal{F}_j)_{j\in J}, \mathcal{F}\Big) = \bigcup_{F\in\mathcal{F}} \bigcap_{j\in F} \mathcal{F}_j,$
2. *if $(\mathcal{G}_l)_{l\in L}$ is a family of filters on J, and $\mathcal{F} = \bigcap_{l\in L} \mathcal{G}_l$, then*
$$\mathcal{D}\Big((\mathcal{F}_j)_{j\in J}, \mathcal{F}\Big) = \bigcap_{l\in L} \mathcal{D}\Big((\mathcal{F}_j)_{j\in J}, \mathcal{G}_l\Big),$$
3. $\mathcal{D}\Big((\mathcal{F}_j)_{j\in J}, \mathcal{F}\Big) = \bigcap_{(\mathcal{U}_j)_{j\in J} \in \prod_{j\in J} \mathbf{U}(\mathcal{F}_j)} \mathcal{D}\Big((\mathcal{U}_j)_{j\in J}, \mathcal{F}\Big),$
4. *if all filters involved are ultrafilters, then so is $\mathcal{D}\Big((\mathcal{F}_j)_{j\in J}, \mathcal{F}\Big)$.*

Proof We only prove the third property, leaving the remaining ones to the reader. One inclusion is clear; to show the other one, suppose that $A \notin \mathcal{D}\Big((\mathcal{F}_j)_j, \mathcal{F}\Big)$. For any $j \in J$, we now choose an ultrafilter in the following way:

$$\begin{cases} \text{if } A \notin \mathcal{F}_j & \text{choose } \mathcal{U}_j \in \mathbf{U}(\mathcal{F}_j) \text{ such that } A \notin \mathcal{U}_j, \\ \text{if } A \in \mathcal{F}_j & \text{choose } \mathcal{U}_j \in \mathbf{U}(\mathcal{F}_j) \text{ arbitrary.} \end{cases}$$

Then, since $A \notin \mathcal{D}\Big((\mathcal{F}_j)_j, \mathcal{F}\Big)$, it follows from the first property that, for any $F \in \mathcal{F}$, there exists $j \in F$ such that $A \notin \mathcal{F}_j$ and, hence, $A \notin \mathcal{U}_j$. Consequently, $A \notin \mathcal{D}\Big((\mathcal{U}_j)_j, \mathcal{F}\Big)$.

□

1.3 APPROACH SYSTEMS

The following structure which we introduce is a kind of localization of the notion of metric. In each point of the space X we give a collection of functions from X to $[0, \infty]$, called local distances, each of which measures a distance from the given point to any other point of the space. Here too, in order to give the precise definitions, we require some preliminary concepts.

A subset $\mathcal{A}$ of functions from a given set X to $[0, \infty]$ is called an *ideal* in $[0, \infty]^X$ if it is closed under the operation of taking finite suprema, and if it closed under the operation of taking smaller functions. In other words, if it is an ideal (dual filter) in the lattice $[0, \infty]^X$ in the order-theoretic sense, when $[0, \infty]^X$ is equipped with the pointwise order (see e.g. Birkhoff [14]).

Given a collection of functions $\mathcal{A} \subset [0, \infty]^X$ and a function $\varphi \in [0, \infty]^X$, we will say that φ is *dominated by* $\mathcal{A}$, or that $\mathcal{A}$ *dominates* φ, if

$$\forall \epsilon > 0,\ \forall \omega < \infty : \ \exists \varphi_\epsilon^\omega \in \mathcal{A} \text{ such that } \varphi \wedge \omega \leq \varphi_\epsilon^\omega + \epsilon.$$

We will then also say that the family $(\varphi_\epsilon^\omega)_{\epsilon > 0, \omega < \infty}$ dominates φ.

Further we will say that a collection of functions $\mathcal{A} \subset [0, \infty]^X$ is *saturated,* if any function which is dominated by $\mathcal{A}$ already belongs to $\mathcal{A}$.

1.3.1 Definition A collection of ideals $(\mathcal{A}(x))_{x \in X}$ in $[0, \infty]^X$, indexed by the points of X, is called an *approach system (on X)* if for all $x \in X$ the following properties hold:

(A1) $\forall \varphi \in \mathcal{A}(x) : \ \varphi(x) = 0,$

(A2) $\mathcal{A}(x)$ is saturated,

(A3) $\forall \varphi \in \mathcal{A}(x),\ \forall \epsilon > 0,\ \forall \omega < \infty\ \exists (\varphi_z)_{z \in X} \in \prod_{z \in X} \mathcal{A}(z)$ such that

$$\forall z, y \in X : \ \varphi(y) \wedge \omega \leq \varphi_x(z) + \varphi_z(y) + \epsilon.$$

For any $x \in X$, a function in $\mathcal{A}(x)$ is called a *local distance* (in x). Note that (A2) implies that $\mathcal{A}(x)$ is closed under the operation of taking smaller functions and that (A1) then implies that the constant function 0 is in $\mathcal{A}(x)$. So the only condition of ideal which, on top of (A1)–(A3), we need to impose explicitly is the one which stipulates that $\mathcal{A}(x)$ must be closed under the formation of finite suprema.

The value $\varphi(y)$ of a local distance $\varphi \in \mathcal{A}(x)$ at a point $y \in X$ is interpreted as "the distance from x to y according to φ". The set of local distances in a point can be compared to the set of neighborhoods of a point in a topological space. Each neighborhood determines its own set of points which are considered close by (in the neighborhood of) the given point. In the same way each local distance makes its own measurement of the distance other points in the space are away from the given point.

Often one can determine collections $\mathcal{B}(x)$, $x \in X$, which would be natural candidates to form an approach system, but not all required properties are fulfilled. In particular property (A2) is not often automatically fulfilled. To handle this we introduce a type of base for approach systems. This too can be compared to the notion of a basis for the neighborhoods in a topological space. We recall that a subset $\mathcal{B}$ of $[0,\infty]^X$ is called an *ideal basis* in $[0,\infty]^X$ if, for any $\alpha, \beta \in \mathcal{B}$, there exists $\gamma \in \mathcal{B}$ such that $\alpha \vee \beta \leq \gamma$. In other words, if $\mathcal{B}$ is an ideal basis (dual filter basis) in the order-theoretic sense (always considering $[0,\infty]^X$ to be equipped with the pointwise order).

1.3.2 Definition A collection of ideal bases $(\mathcal{B}(x))_{x\in X}$ in $[0,\infty]^X$ is called an *approach basis* if, for all $x \in X$, the following properties are satisfied:

(B1) $\forall \varphi \in \mathcal{B}(x): \ \varphi(x) = 0$,

(B2) $\forall \varphi \in \mathcal{B}(x), \forall \epsilon > 0, \ \forall \omega < \infty \ \exists (\varphi_z)_{z\in X} \in \prod_{z\in X} \mathcal{B}(z)$ such that

$$\forall z, y \in X: \ \varphi(y) \wedge \omega \leq \varphi_x(z) + \varphi_z(y) + \epsilon.$$

Notice that (B1) is actually (A1) and that (B2) is actually (A3). Hence an approach system is also an approach basis, and any result for approach bases will also hold for approach systems.

In order to derive the set of all local distances from an approach basis we will also require the following *saturation operation*. Given a subset $\mathcal{B} \subset [0,\infty]^X$ we define

$$\widehat{\mathcal{B}} \doteqdot \left\{\varphi \in [0,\infty]^X \mid \mathcal{B} \text{ dominates } \varphi\right\}.$$

We call $\widehat{\mathcal{B}}$ the *saturation* of $\mathcal{B}$.

1.3.3 Definition A collection of ideal bases $(\mathcal{B}(x))_{x\in X}$ is called a *basis for an approach system* $(\mathcal{A}(x))_{x\in X}$, if for all $x \in X$, $\mathcal{A}(x)$ equals the saturation of $\mathcal{B}(x)$, i.e. $\mathcal{A}(x) = \widehat{\mathcal{B}(x)}$. In this case we also say that $(\mathcal{B}(x))_{x\in X}$ *generates* $(\mathcal{A}(x))_{x\in X}$ or that $(\mathcal{A}(x))_{x\in X}$ is *generated by* $(\mathcal{B}(x))_{x\in X}$.

1.3.4 Proposition *If $(\mathcal{B}(x))_{x\in X}$ is an approach basis, then $\left(\widehat{\mathcal{B}(x)}\right)_{x\in X}$ is an approach system with $(\mathcal{B}(x))_{x\in X}$ as basis.*

Proof (A1) is trivial. To prove (A2) let $x \in X$ and $\varphi \in [0,\infty]^X$ be such that, for all $\omega < \infty$ and $\epsilon > 0$, there exists $\psi \in \widehat{\mathcal{B}(x)}$ such that $\varphi \wedge \omega \leq \psi + \frac{\epsilon}{2}$. If we choose $\tau \in \mathcal{B}(x)$ such that $\psi \wedge \omega \leq \tau + \frac{\epsilon}{2}$, then it follows that $\varphi \wedge \omega \leq \tau + \epsilon$. To prove (A3) let $\varphi \in \widehat{\mathcal{B}(x)}$ and let $\omega < \infty$ and $\epsilon > 0$. Choose $\psi \in \mathcal{B}(x)$ such that $\varphi \wedge \omega \leq \psi + \frac{\epsilon}{2}$ and then choose $(\psi_z)_z \in \prod\limits_{z\in X} \mathcal{B}(z)$ such that, for all $z, y \in X$,

$$\psi(x) \wedge \omega \leq \psi_x(z) + \psi_z(y) + \frac{\epsilon}{2}.$$

Then it follows that, for all $z, y \in X$,

$$\begin{aligned}\varphi(x) \wedge \omega &\leq \left(\psi(x) + \frac{\epsilon}{2}\right) \wedge \left(\omega + \frac{\epsilon}{2}\right) \\ &= \psi(x) \wedge \omega + \frac{\epsilon}{2} \\ &\leq \psi_x(z) + \psi_z(y) + \epsilon.\end{aligned}$$

□

1.3.5 Proposition *If $(\mathcal{B}(x))_{x\in X}$ is a basis for an approach system $(\mathcal{A}(x))_{x\in X}$, then it is an approach basis.*

Proof Clearly, for any $x \in X$, $\mathcal{B}(x)$ is a basis for an ideal and (B1) is fulfilled. To prove (B2) let $\psi \in \mathcal{B}(x)$ and let $\omega < \infty$ and $\epsilon > 0$. Choose $(\varphi_z)_z \in \prod\limits_{z\in X} \mathcal{A}(z)$ such that, for all $z, y \in X$,

$$\psi(y) \wedge \omega \leq \varphi_x(z) + \varphi_z(y) + \frac{\epsilon}{2}.$$

For each $\varphi_z \in \mathcal{A}(z)$, choose $\psi_z \in \mathcal{B}(z)$ such that

$$\varphi_z \wedge \omega \leq \psi_z + \frac{\epsilon}{4}.$$

Then it follows that, for all $z, y \in X$,

$$\begin{aligned}\psi(y) \wedge \omega &\leq \left(\varphi_x(z) + \varphi_z(y) + \frac{\epsilon}{2}\right) \wedge \left(\omega + \frac{\epsilon}{2}\right) \\ &\leq \varphi_x(z) \wedge \omega + \varphi_z(y) \wedge \omega + \frac{\epsilon}{2} \\ &\leq \psi_x(z) + \psi_z(y) + \epsilon.\end{aligned}$$

□

1.4 GAUGES

The type of structure which we introduce in this section should be compared with the definition of a uniform space via a family of p-metrics, called a *uniform gauge*. We refer to Appendix B for definitions and notations in the realm of metric spaces.

Given a collection $\mathcal{D} \subset pq\mathrm{M}^\infty(X)$ and an ∞pq-metric $d \in pq\mathrm{M}^\infty(X)$, we will say that d is *dominated by* $\mathcal{D}$, or that $\mathcal{D}$ *dominates* d, if

$$\forall x \in X,\ \forall \epsilon > 0,\ \forall \omega < \infty : \ \exists d_x^{\epsilon,\omega} \in \mathcal{D} \text{ such that } d(x,\cdot) \wedge \omega \leq d_x^{\epsilon,\omega}(x,\cdot) + \epsilon.$$

We will then also say that the family $\left(d_x^{\epsilon,\omega}\right)_{x \in X, \epsilon > 0, \omega < \infty}$ dominates d.

Further we will say that a collection of ∞pq-metrics $\mathcal{D}$ is *saturated*, if any ∞pq-metric d which is dominated by $\mathcal{D}$ already belongs to $\mathcal{D}$.

1.4.1 Definition A subset $\mathcal{G}$ of $pq\mathrm{M}^\infty(X)$ is called a *gauge* if it is an ideal in $pq\mathrm{M}^\infty(X)$ which fulfills the following property:

(G1) $\mathcal{G}$ is saturated.

As was the case for approach systems, here too, it regularly happens that one has a collection of ∞pq-metrics which would be a natural candidate to form a gauge but not all conditions are fulfilled. The following type of collection will often be encountered. We recall that a subset $\mathcal{H}$ of $pq\mathrm{M}^\infty(X)$ is an ideal basis in $pq\mathrm{M}^\infty(X)$ if for any $d, e \in \mathcal{H}$ there exists $c \in \mathcal{H}$ such that $d \vee e \leq c$.

1.4.2 Definition A subset $\mathcal{H}$ of $pq\mathrm{M}^\infty(X)$ is called a *gauge basis* if it is an ideal basis in $pq\mathrm{M}^\infty(X)$.

By definition, a gauge is also a gauge basis, and similarly to the situation for approach systems and approach bases, here, too, any result shown to hold for gauge bases will also hold for gauges.

In order to derive the gauge from a gauge basis we will also require a *saturation operation* which is perfectly similar to the one for approach systems. Given a subset $\mathcal{D} \subset pq\mathrm{M}^{\infty}(X)$ we define

$$\tilde{\mathcal{D}} \doteqdot \{d \in pq\mathrm{M}^{\infty}(X) \mid \mathcal{D} \text{ dominates } d\}.$$

We call $\tilde{\mathcal{D}}$ the *saturation* of $\mathcal{D}$.

1.4.3 Definition An ideal basis $\mathcal{H}$ in $pq\mathrm{M}^{\infty}(X)$ is said to be a *basis for a gauge* $\mathcal{G}$ if $\tilde{\mathcal{H}} = \mathcal{G}$. In this case we also say that $\mathcal{H}$ *generates* $\mathcal{G}$ or that $\mathcal{G}$ is *generated by* $\mathcal{H}$.

As the following example shows, it is possible for a gauge to be generated by different gauge bases.

1.4.4 Example Consider the real line $\mathbb{R}$ with its usual metric d_E. Then, for any collection of real numbers $A \subset [0, 1]$ such that $\sup A = 1$,

$$\mathcal{H}' \doteqdot \{ad_E \mid a \in A\}$$

is a gauge basis which generates the same gauge as $\mathcal{H} \doteqdot \{d_E\}$.

1.4.5 Proposition *If $\mathcal{H}$ is a gauge basis, then $\tilde{\mathcal{H}}$ is a gauge with $\mathcal{H}$ as basis.*

Proof This goes along the same lines as 1.3.4. □

1.4.6 Proposition *If $\mathcal{H}$ is a basis for a gauge $\mathcal{G}$, then it is a gauge basis.*

Proof This goes along the same lines as 1.3.5. □

1.5 TOWERS

The next structure which we introduce is an ordered family of pretopologies on X, indexed by the real numbers in $[0, \infty]$, and fulfilling certain coherence conditions. The axioms are presented here in terms of closures, but it is clear of course that equivalent sets of axioms can be formulated in terms of any of the defining structures of pretopological spaces, in particular neighborhood systems or convergence. In the recent book by Dikranjan and Tholen [29] it is shown that this axiomatization of approach spaces presents some interesting features as a closure operator.

1.5.1 Definition A family of functions

$$\mathfrak{t}_\epsilon : 2^X \longrightarrow 2^X \qquad \epsilon \in \mathbb{R}^+,$$

is called a *tower (on X)*, if it satisfies the following properties:

(T1) $\forall A \in 2^X,\ \forall \epsilon \in \mathbb{R}^+ : \ A \subset \mathfrak{t}_\epsilon(A)$,

(T2) $\forall \epsilon \in \mathbb{R}^+ : \ \mathfrak{t}_\epsilon(\emptyset) = \emptyset$,

(T3) $\forall A, B \in 2^X,\ \forall \epsilon \in \mathbb{R}^+ : \ \mathfrak{t}_\epsilon(A \cup B) = \mathfrak{t}_\epsilon(A) \cup \mathfrak{t}_\epsilon(B)$,

(T4) $\forall A \in 2^X,\ \forall \epsilon, \gamma \in \mathbb{R}^+ : \ \mathfrak{t}_\epsilon(\mathfrak{t}_\gamma(A)) \subset \mathfrak{t}_{\epsilon+\gamma}(A)$,

(T5) $\forall A \in 2^X,\ \forall \epsilon \in \mathbb{R}^+ : \ \mathfrak{t}_\epsilon(A) = \bigcap_{\epsilon<\gamma} \mathfrak{t}_\gamma(A)$.

Notice that by (T3) and (T5) we have

$$\forall A \subset B \subset X,\ \forall \alpha \leq \beta \in \mathbb{R}^+ : \quad \mathfrak{t}_\alpha(A) \subset \mathfrak{t}_\beta(B).$$

We recall that a *pretopology* on a set X (see also Appendix B) is determined by an operator

$$\mathrm{cl} : 2^X \longrightarrow 2^X$$

which fulfills the properties: (1) $A \subset \mathrm{cl}(A)$, (2) $\mathrm{cl}(\emptyset) = \emptyset$, and (3) $\mathrm{cl}(A \cup B) = \mathrm{cl}(A) \cup \mathrm{cl}(B)$, for all $A, B \in 2^X$. This operator is then called a *pretopological closure operator*. A set X equipped with a pretopology is called a *pretopological space*. Pretopologies and pretopological spaces and continuous maps between them were introduced by Choquet in [25]. They form the objects and morphisms of a topological construct (see Appendices A and B).

1.5.2 Proposition *The following are equivalent:*

1. $(\mathfrak{t}_\epsilon)_{\epsilon \in \mathbb{R}^+}$ *is a tower,*

2. for all $\epsilon \in \mathbb{R}^+$, $\mathfrak{t}_\epsilon$ *is a pretopological closure operator, and (T4) and (T5) are fulfilled,*

3. $\mathfrak{t}_0$ *is a topological closure operator, for all* $\epsilon > 0$, $\mathfrak{t}_\epsilon$ *is a pretopological closure operator, and (T4) and (T5) are fulfilled.*

Proof Properties (T1), (T2), and (T3) constitute the definition of a pretopological closure operator.

□

1.6 HULL OPERATORS

The type of structure which we consider in this section is a notion of hull operator for real-valued functions, and it is to be compared with well-known operations such as lower or upper semicontinuous regularization, convex regularization, or nonexpansive regularization of functions, as e.g. in Bourbaki [17] and in Singer [111]. It should be pointed out, however, that of the above examples, only lower semicontinuous regularization and nonexpansive regularization will turn out to be hull operators in the sense of our definition. Convex regularization fails to satisfy property (H2) and upper semicontinuous regularization fails to fulfill (H1) and (H2). For precise details on these examples we refer to the second and third chapters.

1.6.1 Definition A function

$$h : [0,\infty]^X \longrightarrow [0,\infty]^X$$

is called a *hull operator (on X)* if it satisfies the following properties:

(H1) $\forall \mu \in [0,\infty]^X : \quad h(\mu) \leq \mu$,

(H2) $\forall \mu, \nu \in [0,\infty]^X : \quad h(\mu \wedge \nu) = h(\mu) \wedge h(\nu)$,

(H3) $\forall \mu \in [0,\infty]^X : \quad h(h(\mu)) = h(\mu)$,

(H4) $\forall \mu \in [0,\infty]^X, \ \forall \alpha \in [0,\infty] : \quad h(\mu + \alpha) = h(\mu) + \alpha$.

1.6.2 Proposition *If* $h : [0,\infty]^X \longrightarrow [0,\infty]^X$ *is a hull operator, then the following hold:*

1. $\forall \alpha$ *constant*: $h(\alpha) = \alpha$,

2. $\forall \mu \in [0,\infty]^X, \ \forall \alpha \in \left[0, \inf_{x \in X} \mu(x)\right]: \ h(\mu - \alpha) = h(\mu) - \alpha$,

3. $\forall \mu \in [0,\infty]^X, \ \forall \alpha \in [0,\infty]: \ h((\mu - \alpha) \vee 0) \geq (h(\mu) - \alpha) \vee 0$.

Proof The first property follows from (H4) and (H1) letting $\mu = 0$. The second property follows from (H4) since

$$\begin{aligned} h(\mu) &= h(\mu - \alpha + \alpha) \\ &= h(\mu - \alpha) + \alpha. \end{aligned}$$

The third property follows from the second one and the fact that $(\mu - \alpha) \vee 0 = \mu \vee \alpha - \alpha$.

□

In what follows, especially in our investigations concerning hull operators, we will sometimes need to restrict our attention to bounded functions in $[0,\infty]^X$. We will denote the collection of these functions by $[0,\infty]_b^X$.

1.6.3 Proposition *A hull operator h on X is completely determined by its restriction to $[0,\infty]_b^X$. In particular, for any $\mu \in [0,\infty]^X$ and any set $K \subset [0,\infty[$ such that $\sup K = \infty$, we have*

$$h(\mu) = \sup_{\alpha\in K} h(\mu \wedge \alpha).$$

Proof This follows from the fact that, by (H2) and 1.6.2, we have

$$\sup_{\alpha\in K} h(\mu \wedge \alpha) = \sup_{\alpha\in K} h(\mu) \wedge \alpha.$$

□

1.6.4 Proposition *If*

$$h : [0,\infty]_b^X \longrightarrow [0,\infty]_b^X$$

fulfills the properties

(H1b) $\forall \mu \in [0,\infty]_b^X : \ h(\mu) \leq \mu$,
(H2b) $\forall \mu, \nu \in [0,\infty]_b^X : \ h(\mu \wedge \nu) = h(\mu) \wedge h(\nu)$,
(H3b) $\forall \mu \in [0,\infty]_b^X : \ h(h(\mu)) = h(\mu)$,
(H4b) $\forall \mu \in [0,\infty]_b^X, \ \forall \alpha \in \mathbb{R}^+ : \ h(\mu + \alpha) = h(\mu) + \alpha$,

and we define

$$h^* : [0,\infty]^X \longrightarrow [0,\infty]^X$$

by

$$h^*(\mu) \doteqdot \sup_{\alpha<\infty} h(\mu \wedge \alpha),$$

then h^* is the unique hull operator on X, the restriction to $[0,\infty]_b^X$ of which coincides with h.

Proof (H1) and (H2) are immediate. To prove (H3) let $\mu \in [0,\infty]^X$; then

$$\begin{aligned} h^*(h^*(\mu)) &= \sup_{\alpha<\infty} h\left(\left(\sup_{\beta<\infty} h(\mu\wedge\beta)\right)\wedge\alpha\right) \\ &= \sup_{\alpha<\infty} h\left(\sup_{\beta<\infty} (h(\mu\wedge\alpha)\wedge\beta)\right) \\ &= \sup_{\alpha<\infty} h(\mu\wedge\alpha) \\ &= h^*(\mu). \end{aligned}$$

(H4) follows from the observation that, for any $\mu \in [0,\infty]^X$ and any $\alpha < \infty$, we have

$$\begin{aligned} h^*(\mu+\alpha) &= \sup_{\alpha\leq\beta<\infty} h(\mu\wedge(\beta-\alpha)+\alpha) \\ &= \sup_{\alpha\leq\beta<\infty} h(\mu\wedge(\beta-\alpha))+\alpha \\ &= h^*(\mu)+\alpha. \end{aligned}$$

Uniqueness finally follows from 1.6.3.

□

In the sequel we will also require the following type of characteristic maps. For any $A \subset X$, we define

$$\theta_A : X \longrightarrow [0,\infty] : x \longrightarrow \begin{cases} 0 & x \in A, \\ \infty & x \notin A. \end{cases}$$

We will call this map the *indicator of A*. The set of all such indicators on X will be denoted by $Ind(X)$.

Further we will require to approximate functions in $[0,\infty]_b^X$ uniformly from below by functions taking only a finite number of values. We will formalize the procedure in the following way.

Let us denote the set of all functions attaining only a finite number of values in $\mathbb{R}^+$ by $Fin(X)$. Clearly, any function in $Fin(X)$ can be written as $\inf_{i=1}^{n}(a_i + \theta_{A_i})$, where $(A_i)_{i=1}^n$ is a partitioning of X.

Given $\mu \in [0,\infty]_b^X$, we will say that the family $(\mu_\epsilon)_{\epsilon>0}$ in $Fin(X)$ is a *development of* μ if

$$\forall \epsilon > 0 : \ \mu_\epsilon \leq \mu \leq \mu_\epsilon + \epsilon.$$

1.6.5 Proposition *A hull operator h on X is completely determined by its restriction to $Ind(X)$. In particular, for any $\mu \in [0,\infty]_b^X$ and any development $\left(\mu_\epsilon \doteq \inf_{i=1}^{n(\epsilon)} (m_i^\epsilon + \theta_{M_i^\epsilon})\right)_{\epsilon>0}$ of μ, we have*

$$h(\mu) = \sup_{\epsilon>0} \left(\inf_{i=1}^{n(\epsilon)} (m_i^\epsilon + h(\theta_{M_i^\epsilon})) \right).$$

Proof Applying (H2) and (H4) it follows that, for any $\epsilon > 0$, we have

$$h(\mu_\epsilon) \leq h(\mu) \leq h(\mu_\epsilon) + \epsilon,$$

and hence, once again applying (H2) and (H4), it follows that

$$\begin{aligned} h(\mu) &= \sup_{\epsilon>0} h(\mu_\epsilon) \\ &= \sup_{\epsilon>0} \left(\inf_{i=1}^{n(\epsilon)} (m_i^\epsilon + h(\theta_{M_i^\epsilon})) \right). \end{aligned}$$

□

1.6.6 Corollary *If h is a hull operator on X, $\mu \in [0,\infty]^X$, and, for each $\omega < \infty$, $\left(\inf_{i=1}^{n(\epsilon,\omega)} \left(m_i^{\epsilon,\omega} + \theta_{M_i^{\epsilon,\omega}}\right)\right)_{\epsilon>0}$ is a development of $\mu \wedge \omega$, then we have*

$$h(\mu) = \sup_{\omega<\infty} \sup_{\epsilon>0} \left(\inf_{i=1}^{n(\epsilon,\omega)} \left(m_i^{\epsilon,\omega} + h\left(\theta_{M_i^{\epsilon,\omega}}\right)\right) \right).$$

□

1.7 REGULAR FUNCTION FRAMES

The final structure which we consider is that of the collection of functions, each of which is equal to its own hull in the sense of the previous definition. Referring to the examples of the previous section, these structures then are to be compared with collections of lower semicontinuous functions or collections of nonexpansive functions.

1.7.1 Definition A collection of functions $\mathcal{R} \subset [0, \infty]^X$ is called a *regular function frame (on X)*, if it satisfies the following properties:

(R1) $\forall \mathcal{S} \subset \mathcal{R} : \bigvee \mathcal{S} \in \mathcal{R}$,

(R2) $\forall \mu, \nu \in \mathcal{R} : \mu \wedge \nu \in \mathcal{R}$,

(R3) $\forall \mu \in \mathcal{R}, \forall \alpha \in [0, \infty] : \mu + \alpha \in \mathcal{R}$,

(R4) $\forall \mu \in \mathcal{R}, \forall \alpha \in \left[0, \inf\limits_{x \in X} \mu(x)\right] : \mu - \alpha \in \mathcal{R}$.

The members of $\mathcal{R}$ are called *regular functions.*

1.7.2 Proposition *If $\mathcal{R} \subset [0, \infty]^X$ is a regular function frame, then the following hold:*

1. *$\mathcal{R}$ contains all constant functions,*
2. $\forall \mu \in \mathcal{R}, \forall \alpha \in [0, \infty] : (\mu - \alpha) \vee 0 \in \mathcal{R}$.

Proof Letting $\mathcal{S} = \emptyset$ in (R1) it follows that $0 \in \mathcal{R}$ and then the first property follows from (R3) letting $\mu = 0$. The second property follows from the first one, (R1), and (R4).

□

1.8 APPROACH SPACES

In spite of the fact that all the concepts which were defined in 1.1 to 1.7 are both conceptually and technically very different from each other, they are all equivalent. It is the purpose of the present section to prove this. Not only will we prove that one type of structure unambiguously determines a unique structure of each of the other types, but we will also give several precise formulas for going from one structure to another. A structure derived from another one by such a transition will be referred to as an *associated structure*; thus we will speak, for example, of the gauge associated with a distance or of the associated gauge and so on.

Since we have seven different types of structures there are no less than 42 possible transitions. Of course we will not attempt to describe them all but we will restrict ourselves to those which are needed to prove the equivalence of the different structures and to those which are interesting and necessary for later. Each theorem giving such a transition will be marked with an indication of its precise nature, e.g. as (**L** $\rightrightarrows$ **R**) for the transition from limit operators to regular function frames. (See also the tables on pp. 55–58.)

The plan of this section is as follows. First, we will give the following transitions, from which it will result that all structures are equivalent with each other:

$$\begin{array}{ccccccc} & & \mathrm{t} & & & & \\ & & \Updownarrow & & & & \\ \lambda & \Leftrightarrow & \delta & \Leftrightarrow & \mathcal{G} & \Leftrightarrow & \mathcal{A} \\ & & \Updownarrow & & & & \\ & & h & & & & \\ & & \Updownarrow & & & & \\ & & \mathcal{R} & & & & \end{array}$$

Second, we will give explicit noncircuitous formulas for the following transitions:

$$\begin{array}{ccccc} \delta & \Leftrightarrow & \mathcal{A} & \Rightarrow & \lambda \\ & & \Updownarrow & & \Uparrow \\ & & h & \Leftarrow & \mathcal{G} \end{array}$$

1.8.1 Theorem (D $\rightrightarrows$ L) *If $\delta : X \times 2^X \longrightarrow [0, \infty]$ is a distance on X, then the function*

$$\lambda\colon \mathbf{F}(X) \longrightarrow [0,\infty]^X : \mathcal{F} \longrightarrow \sup_{U \in \mathrm{sec}(\mathcal{F})} \delta_U$$

is a limit operator on X. Moreover, for any $x \in X$ and $A \in 2^X$, we have

$$\delta(x, A) = \inf_{\mathcal{U} \in \mathbf{U}(A)} \lambda\mathcal{U}(x).$$

Proof (L1) follows from (D1), and (L2) follows from the fact that $\mathcal{F} \subset \mathcal{G}$ implies $\mathbf{U}(\mathcal{G}) \subset \mathbf{U}(\mathcal{F})$. To prove (L3), let $(\mathcal{F}_j)_{j\in J}$ be a family of filters on X. Then, upon noticing that $\sec\left(\bigcap_{j\in J}\mathcal{F}_j\right) = \bigcup_{j\in J}\sec\mathcal{F}_j$, it follows that we have

$$\begin{aligned}\lambda\left(\bigcap_{j\in J}\mathcal{F}_j\right) &= \sup_{U\in\sec(\bigcap_{j\in J}\mathcal{F}_j)} \delta_U \\ &= \sup_{U\in\bigcup_{j\in J}\sec\mathcal{F}_j} \delta_U \\ &= \sup_{j\in J}\sup_{U\in\sec(\mathcal{F}_j)} \delta_U \\ &= \sup_{j\in J}\lambda(\mathcal{F}_j).\end{aligned}$$

To prove (L4), let $\mathcal{F}\in\mathbf{F}(X)$ and let $(\mathcal{S}(y))_{y\in X}$ be a selection of filters on X. Next let

$$\epsilon \doteqdot \sup_{y\in X}\lambda(\mathcal{S}(y))(y).$$

First suppose that all filters involved are ultrafilters. For any $D\in\mathcal{D}(\mathcal{S},\mathcal{F})$, there exists an $F\in\mathcal{F}$ such that, for all $y\in F$, $D\in\mathcal{S}(y)$. Consequently,

$$\delta(y,D)\leq\lambda(\mathcal{S}(y))(y)\leq\epsilon.$$

This proves that $D^{(\epsilon)}\in\mathcal{F}$ and then it follows from (D4) that

$$\begin{aligned}\delta_D &\leq \delta_{D^{(\epsilon)}}+\epsilon \\ &\leq \sup_{F\in\mathcal{F}}\delta_F+\epsilon \\ &\leq \lambda\mathcal{F}+\epsilon.\end{aligned}$$

By the arbitrariness of $D\in\mathcal{D}(\mathcal{S},\mathcal{F})$ it follows that

$$\begin{aligned}\lambda\mathcal{D}(\mathcal{S},\mathcal{F}) &= \sup_{D\in\mathcal{D}(\mathcal{S},\mathcal{F})}\delta_D \\ &\leq \lambda\mathcal{F}+\epsilon.\end{aligned}$$

Second we consider the case where all filters involved are arbitrary. For each selection $\mathcal{R}\in\prod_{y\in X}\mathbf{U}(\mathcal{S}(y))$, let

$$\epsilon_{\mathcal{R}} \doteqdot \sup_{y\in X}\lambda(\mathcal{R}(y))(y).$$

A straightforward verification shows that

$$\epsilon = \sup_{\mathcal{R} \in \prod_{y \in X} \mathbf{U}(\mathcal{S}(y))} \epsilon_{\mathcal{R}}.$$

From the result for ultrafilters it then follows that, for any $\mathcal{R} \in \prod_{y \in X} \mathbf{U}(\mathcal{S}(y))$ and $\mathcal{U} \in \mathbf{U}(\mathcal{F})$,

$$\lambda \mathcal{D}(\mathcal{R}, \mathcal{U}) \leq \lambda \mathcal{U} + \epsilon_{\mathcal{R}}.$$

It then follows from (L3) that

$$\begin{aligned} \lambda \mathcal{D}(\mathcal{S}, \mathcal{F}) &= \sup_{\mathcal{R} \in \prod_{y \in X} \mathbf{U}(\mathcal{S}(y))} \sup_{\mathcal{U} \in \mathbf{U}(\mathcal{F})} \lambda \mathcal{D}(\mathcal{R}, \mathcal{U}) \\ &\leq \sup_{\mathcal{R} \in \prod_{y \in X} \mathbf{U}(\mathcal{S}(y))} \sup_{\mathcal{U} \in \mathbf{U}(\mathcal{F})} (\lambda \mathcal{U} + \epsilon_{\mathcal{R}}) \\ &= \lambda \mathcal{F} + \epsilon. \end{aligned}$$

To prove the final claim of the theorem, first notice that one inequality is clear. To prove the other one, let $x \in X$ and $A \in 2^X$. It follows from the definition of λ, and upon applying complete distributivity, that

$$\inf_{\mathcal{U} \in \mathbf{U}(A)} \lambda \mathcal{U} = \sup_{\theta \in \prod_{\mathcal{U} \in \mathbf{U}(A)} \mathcal{U}} \inf_{\mathcal{U} \in \mathbf{U}(A)} \delta_{\theta(\mathcal{U})}.$$

Now, by 1.2.2, for each $\theta \in \prod_{\mathcal{U} \in \mathbf{U}(A)} \mathcal{U}$, we can find a finite subset $\mathbf{U}_\theta \subset \mathbf{U}(A)$ such that $A \subset \bigcup_{\mathcal{U} \in \mathbf{U}_\theta} \theta(\mathcal{U})$. Consequently it follows from (D3) that

$$\begin{aligned} \inf_{\mathcal{U} \in \mathbf{U}(A)} \lambda \mathcal{U} &\leq \sup_{\theta \in \prod_{\mathcal{U} \in \mathbf{U}(A)} \mathcal{U}} \delta_{\left(\bigcup_{\mathcal{U} \in \mathbf{U}_\theta} \theta(\mathcal{U}) \right)} \\ &\leq \delta_A, \end{aligned}$$

which proves the remaining inequality.

□

1.8.2 Theorem (L $\rightrightarrows$ D) *If $\lambda : \mathbf{F}(X) \longrightarrow [0, \infty]^X$ is a limit operator on X, then the function*

$$\delta : X \times 2^X \longrightarrow [0, \infty] : (x, A) \longrightarrow \inf_{\mathcal{U} \in \mathbf{U}(A)} \lambda \mathcal{U}(x)$$

is a distance on X. *Moreover, for any* $\mathcal{F} \in \mathbf{F}(X)$ *and* $x \in X$, *we have*

$$\lambda\mathcal{F}(x) = \sup_{U \in \mathrm{sec}(\mathcal{F})} \delta(x, U).$$

Proof (D1) follows from (L1), (D2) is trivial, and (D3) follows from the fact that, for any $A, B \in 2^X$, we have $\mathbf{U}(\mathrm{stack} A \cup B) = \mathbf{U}(\mathrm{stack} A) \cup \mathbf{U}(\mathrm{stack} B)$. We will now prove the final claim of the theorem since we will require this in the proof of (D4). Let λ' be defined as

$$\lambda' : \mathbf{F}(X) \longrightarrow [0, \infty]^X : \mathcal{F} \longrightarrow \sup_{U \in \mathrm{sec}\mathcal{F}} \delta_U.$$

Let $\mathcal{U} \in \mathbf{U}(X)$; then first we have

$$\begin{aligned} \lambda'\mathcal{U} &= \sup_{U \in \mathcal{U}} \delta_U \\ &= \sup_{U \in \mathcal{U}} \inf_{\mathcal{W} \in \mathbf{U}(U)} \lambda\mathcal{W} \\ &\leq \lambda\mathcal{U}. \end{aligned}$$

Second, by complete distributivity, we have

$$\begin{aligned} \lambda'\mathcal{U} &= \sup_{U \in \mathcal{U}} \inf_{\mathcal{W} \in \mathbf{U}(U)} \lambda\mathcal{W} \\ &= \inf_{\theta \in \prod_{U \in \mathcal{U}} \mathbf{U}(U)} \sup_{U \in \mathcal{U}} \lambda(\theta(U)). \end{aligned}$$

Furthermore, for any $\theta \in \prod_{U \in \mathcal{U}} \mathbf{U}(U)$ and any $U \in \mathcal{U}$, we have $U \in \theta(U)$ and thus $\bigcap_{U \in \mathcal{U}} \theta(U) \subset \mathcal{U}$. From (L2) it then follows that

$$\lambda\mathcal{U} \leq \lambda\left(\bigcap_{U \in \mathcal{U}} \theta(U)\right).$$

Since this holds for all $\theta \in \prod_{U \in \mathcal{U}} \mathbf{U}(U)$, it follows that

$$\lambda\mathcal{U} \leq \lambda'\mathcal{U}.$$

Consequently λ and λ' coincide on ultrafilters. From the definition of λ' and the fact that λ fulfills (L3), this, however, suffices to conclude that $\lambda = \lambda'$.

In order now to prove (D4), let $A \in 2^X$, $\epsilon \in \mathbb{R}^+$, and $\mathcal{W} \in \mathbf{U}\left(A^{(\epsilon)}\right)$.

Claim: $\forall y \in A^{(\epsilon)}\ \exists\, \mathcal{S}(y) \in \mathbf{U}(A)$ such that $\lambda(\mathcal{S}(y))(y) \leq \epsilon$.

Indeed, if not then, for some $y \in A^{(\epsilon)}$ and for all $\mathcal{U} \in \mathbf{U}(A)$, we have

$$\begin{aligned} \epsilon &< \lambda\mathcal{U}(y) \\ &= \lambda'\mathcal{U}(y) \\ &= \sup_{U \in \mathcal{U}} \delta(y, U). \end{aligned}$$

Consequently, for all $\mathcal{U} \in \mathbf{U}(A)$, we can find $U_{\mathcal{U}} \in \mathcal{U}$ such that $\epsilon < \delta(y, U_{\mathcal{U}})$. By 1.2.2 we can then find $\mathcal{U}_1, \ldots, \mathcal{U}_n \in \mathbf{U}(A)$ such that $A \subset \bigcup_{i=1}^{n} U_{\mathcal{U}_i}$. It then follows from (D3) that

$$\begin{aligned} \epsilon &< \inf_{i=1}^{n} \delta(y, U_{\mathcal{U}_i}) \\ &= \delta\left(y, \bigcup_{i=1}^{n} U_{\mathcal{U}_i}\right) \\ &\leq \delta(y, A), \end{aligned}$$

which is in contradiction with the choice of y. This proves our claim. Consequently, for all $y \in A^{(\epsilon)}$, we can fix some $\mathcal{S}(y) \in \mathbf{U}(A)$ such that $\lambda(\mathcal{S}(y))(y) \leq \epsilon$. For $y \notin A^{(\epsilon)}$ we let $\mathcal{S}(y) \doteq \text{stack}\, y$. Next we let

$$\epsilon' \doteq \sup_{y \in X} \lambda(\mathcal{S}(y))(y).$$

Since $A \in \bigcap_{y \in A^{(\epsilon)}} \mathcal{S}(y)$ it follows that $A \in \mathcal{D}(\mathcal{S}, \mathcal{W})$ and consequently, by 1.2.3, $\mathcal{D}(\mathcal{S}, \mathcal{W}) \in \mathbf{U}(A)$. From the definition of δ and making use of (L4) we then obtain that, for all $x \in X$,

$$\begin{aligned} \delta(x, A) &\leq \lambda(\mathcal{D}(\mathcal{S}, \mathcal{W}))(x) \\ &\leq \lambda\mathcal{W}(x) + \epsilon' \\ &\leq \lambda\mathcal{W}(x) + \epsilon. \end{aligned}$$

Since this holds for all $\mathcal{W} \in \mathbf{U}\left(A^{(\epsilon)}\right)$, it follows that

$$\delta(x, A) \leq \delta\left(x, A^{(\epsilon)}\right) + \epsilon.$$

□

The foregoing two theorems show that distances and limit operators are equivalent concepts. Together with approach systems and gauges they form the most important and useful structures.

We now turn to the relation between distances and gauges.

1.8.3 Theorem (D $\rightrightarrows$ G) *If $\delta : X \times 2^X \longrightarrow [0, \infty]$ is a distance on X, then*

$$\mathcal{G} \doteqdot \left\{ d \in pq\mathbf{M}^\infty(X) \mid \forall A \subset X,\ \forall x \in X : \inf_{a \in A} d(x, a) \leq \delta(x, A) \right\}$$

is a gauge on X.

Proof Let $\mathcal{G}_0 \in 2^{(\mathcal{G})}$, $x \in X$, and $A \subset X$. Then, by complete distributivity, we have that $\inf_{a \in A} \sup_{d \in \mathcal{G}_0} d(x, a) = \sup_{\varphi \in \mathcal{G}_0^A} \inf_{a \in A} \varphi(a)(x, a)$. If we fix $\varphi \in \mathcal{G}_0^A$, then it follows that

$$\begin{aligned} \inf_{a \in A} \varphi(a)(x, a) &= \inf_{d \in \mathcal{G}_0} \inf_{a \in \varphi^{-1}(d)} d(x, a) \\ &\leq \inf_{d \in \mathcal{G}_0} \delta\big(x, \varphi^{-1}(d)\big) \\ &= \delta(x, A). \end{aligned}$$

Hence it follows that $\mathcal{G}$ is closed under the formation of finite suprema. The remaining properties of an ideal are trivial.

If $d \in pq\mathbf{M}^\infty(X)$ is such that, for all $x \in X$, $\epsilon > 0$, and $\omega < \infty$, there exists $d' \in \mathcal{G}$ such that $d(x, \cdot) \wedge \omega \leq d'(x, \cdot) + \epsilon$ then it follows at once that, for any $\epsilon > 0$ and $\omega < \infty$,

$$\begin{aligned} \inf_{a \in A} d(x, a) \wedge \omega &\leq \inf_{a \in A} d'(x, a) + \epsilon \\ &\leq \delta(x, A) + \epsilon. \end{aligned}$$

By the arbitrariness of ϵ and ω this proves that $d \in \mathcal{G}$. Hence $\mathcal{G}$ is saturated. □

In order to be able to work with the gauge associated with a distance, as given by the foregoing theorem, it will be useful to have a specified set of ∞pq-metrics in that gauge. The following proposition provides us with such a collection.

1.8.4 Proposition *If $\delta : X \times 2^X \longrightarrow [0,\infty]$ is a distance on X, then, for any $\zeta < \infty$ and $Z \subset X$, the function*

$$d_Z^\zeta : X \times X \longrightarrow [0,\infty] : (x,y) \longrightarrow (\delta(x,Z) \wedge \zeta - \delta(y,Z) \wedge \zeta) \vee 0$$

is a pq-metric in the gauge associated with δ.

Proof That d_Z^ζ is a *pq*-metric is easily seen. Let $x \in X$ and $A \subset X$. Then it follows from (D4) that

$$\begin{aligned}
\inf_{a\in A} d_Z^\zeta(x,a) &= \inf_{a\in A} (\delta(x,Z) \wedge \zeta - \delta(a,Z) \wedge \zeta) \vee 0 \\
&\leq \left(\delta(x,Z) \wedge \zeta - \sup_{a\in A} \delta(a,Z) \wedge \zeta\right) \vee 0 \\
&\leq \left(\left(\delta(x,A) + \sup_{a\in A} \delta(a,Z)\right) \wedge \zeta - \sup_{a\in A} \delta(a,Z) \wedge \zeta\right) \vee 0 \\
&\leq \left(\delta(x,A) + \sup_{a\in A} \delta(a,Z) \wedge \zeta - \sup_{a\in A} \delta(a,Z) \wedge \zeta\right) \vee 0 \\
&= \delta(x,A).
\end{aligned}$$

By the characterization given in 1.8.3 this proves that d_Z^ζ is indeed a member of the gauge associated with δ.

□

1.8.5 Theorem (G $\rightrightarrows$ D) *If $\mathcal{G} \subset pq\mathrm{M}^\infty(X)$ is a gauge basis on X, then the function*

$$\delta : X \times 2^X \longrightarrow [0,\infty] : (x,A) \longrightarrow \sup_{d\in\mathcal{G}} \inf_{a\in A} d(x,a)$$

is a distance on X.

Proof Verification of (D1), (D2), and (D3) is straightforward. To show (D4) let $x \in X$, $A \subset X$, and $\epsilon > 0$ be fixed. Then, for any $b \in A^{(\epsilon)}$, $d \in \mathcal{G}$, and $\theta > 0$, there exists $a_d \in A$ such that $d(b,a_d) \leq \epsilon + \theta$. Consequently,

$$\begin{aligned}
d(x,a_d) &\leq d(x,b) + d(b,a_d) \\
&\leq d(x,b) + \epsilon + \theta,
\end{aligned}$$

which proves that $\inf_{a\in A} d(x,a) \leq \inf_{b\in A^{(\epsilon)}} d(x,b) + \epsilon + \theta$. Since this holds for all $d \in \mathcal{G}$, it follows that $\delta(x,A) \leq \delta\left(x, A^{(\epsilon)}\right) + \epsilon$.

□

1.8.6 Theorem *If $\delta : X \times 2^X \longrightarrow [0,\infty]$ is a distance on X and $\mathcal{G}$ is the associated gauge, then we have*

$$\delta(x,A) = \sup_{d\in\mathcal{G}} \inf_{a\in A} d(x,a) \qquad \forall x \in X,\ \forall A \subset X.$$

Proof Let $x \in X$ and $A \subset X$. It follows at once from 1.8.3 that $\sup_{d\in\mathcal{G}} \inf_{a\in A} d(x,a) \leq \delta(x,A)$. Making use of 1.8.4, we further obtain

$$\begin{aligned}
\sup_{d\in\mathcal{G}} \inf_{a\in A} d(x,a) &\geq \sup_{\zeta<\infty} \sup_{Z\subset X} \inf_{a\in A} d_Z^{\zeta}(x,a) \\
&\geq \sup_{\zeta<\infty} \inf_{a\in A} d_A^{\zeta}(x,a) \\
&= \sup_{\zeta<\infty} \inf_{a\in A} (\delta(x,A)\wedge\zeta - \delta(a,A)\wedge\zeta) \vee 0 \\
&= \sup_{\zeta<\infty} \inf_{a\in A} \delta(x,A)\wedge\zeta \\
&= \delta(x,A),
\end{aligned}$$

which proves the other inequality.

□

1.8.7 Theorem *If $\mathcal{G} \subset pq\mathrm{M}^{\infty}(X)$ is a gauge on X and δ is the associated distance, then*

$$\mathcal{G} = \left\{ d \in pq\mathrm{M}^{\infty}(X) \mid \forall A \subset X,\ \forall x \in X :\ \inf_{a\in A} d(x,a) \leq \delta(x,A) \right\}.$$

Proof Let us put

$$\mathcal{G}' \doteqdot \left\{ d \in pq\mathrm{M}^{\infty}(X) \mid \forall A \subset X,\ \forall x \in X :\ \inf_{a\in A} d(x,a) \leq \delta(x,A) \right\}.$$

Then we know from 1.8.3 that $\mathcal{G}'$ is the gauge associated with δ. That $\mathcal{G} \subset \mathcal{G}'$ follows at once from 1.8.5. Now suppose that there exists $d' \in \mathcal{G}' \setminus \mathcal{G}$; then d' is not dominated by $\mathcal{G}$ and hence we can find $x \in X$, $\epsilon > 0$, and $\omega < \infty$ such that, for all $d \in \mathcal{G}$,

$$d'(x,\cdot) \wedge \omega \nleq d(x,\cdot) + \epsilon.$$

For each $d \in \mathcal{G}$, let $A(d) \doteqdot \{y \in X \mid d'(x,y) \wedge \omega > d(x,y) + \epsilon\}$. Then it is clear that, for all $d, e \in \mathcal{G}$, $A(d) \cap A(e) = A(d \vee e) \neq \emptyset$. Hence, making use of 1.8.6 we obtain

$$\begin{aligned}
\sup_{d\in\mathcal{G}} \delta(x, A(d)) \wedge \omega &= \sup_{d\in\mathcal{G}} \sup_{e\in\mathcal{G}} \inf_{y\in A(d)} e(x,y) \wedge \omega \\
&\leq \sup_{d\in\mathcal{G}} \sup_{e\in\mathcal{G}} \inf_{y\in A(d\vee e)} (d \vee e)(x,y) \wedge \omega \\
&= \sup_{d\in\mathcal{G}} \inf_{y\in A(d)} d(x,y) \wedge \omega \\
&\leq \sup_{d\in\mathcal{G}} \inf_{y\in A(d)} \big(d'(x,y) \wedge \omega - \epsilon\big) \wedge \omega \\
&\leq \sup_{d\in\mathcal{G}} \inf_{y\in A(d)} d'(x,y) \wedge \omega - \epsilon \\
&\leq \sup_{d\in\mathcal{G}} \sup_{d''\in\mathcal{G}'} \inf_{y\in A(d)} d''(x,y) \wedge \omega - \epsilon \\
&= \sup_{d\in\mathcal{G}} \delta(x, A(d)) \wedge \omega - \epsilon,
\end{aligned}$$

which is a contradiction.

□

The combined results of 1.8.3, 1.8.5, 1.8.6, and 1.8.7 prove that distances and gauges are equivalent systems.

We end our investigation into the relationship between distances and gauges with a result which we will require later, showing that if a collection of ∞pq-metrics produces a formula for a distance as in 1.8.5 then it must be a basis for the associated gauge.

1.8.8 Theorem *If δ is a distance on X and $\mathcal{D} \subset pq\mathrm{M}^{\infty}(X)$ is a collection of ∞pq-metrics such that, for all $x \in X$ and $A \subset X$, we have $\delta(x, A) = \sup_{d\in\mathcal{D}} \inf_{a\in A} d(x,a)$, then $\mathcal{D}$ is a basis for the gauge $\mathcal{G}$ associated with δ.*

Proof It follows from 1.8.7 that $\mathcal{D} \subset \mathcal{G}$. Suppose that there exists $d_0 \in \mathcal{G} \setminus \widetilde{\mathcal{D}}$. Then d_0 is not dominated by $\mathcal{D}$ and hence there exist $x \in X$, $\epsilon > 0$, and $\omega < \infty$ such that, for all $d \in \mathcal{D}$,

$$d_0(x, \cdot) \wedge \omega \not\leq d(x, \cdot) + \epsilon.$$

For each $d \in \mathcal{D}$, let $A(d) \doteqdot \{y \in X \mid d_0(x,y) \wedge \omega > d(x,y) + \epsilon\}$. Then it is clear that, for all $d, e \in \mathcal{G}$, $A(d) \cap A(e) = A(d \vee e) \neq \emptyset$. Hence, making use of 1.8.6, we obtain

$$\begin{aligned}
\sup_{d\in\mathcal{D}} \delta(x, A(d)) \wedge \omega &= \sup_{d\in\mathcal{D}} \sup_{e\in\mathcal{D}} \inf_{y\in A(d)} e(x,y) \wedge \omega \\
&\leq \sup_{d\in\mathcal{D}} \sup_{e\in\mathcal{D}} \inf_{y\in A(d\vee e)} (d\vee e)(x,y) \wedge \omega \\
&= \sup_{d\in\mathcal{D}} \inf_{y\in A(d)} d(x,y) \wedge \omega \\
&\leq \sup_{d\in\mathcal{D}} \inf_{y\in A(d)} (d_0(x,y) \wedge \omega - \epsilon) \wedge \omega \\
&\leq \sup_{d\in\mathcal{D}} \inf_{y\in A(d)} d_0(x,y) \wedge \omega - \epsilon \\
&\leq \sup_{d\in\mathcal{D}} \sup_{e\in\mathcal{D}} \inf_{y\in A(d)} e(x,y) \wedge \omega - \epsilon \\
&= \sup_{d\in\mathcal{D}} \delta(x, A(d)) \wedge \omega - \epsilon,
\end{aligned}$$

which is a contradiction.

□

In the following corollary, given $\zeta < \infty$ and $Z \subset X$, d_Z^ζ stands for the ∞pq-metric defined in 1.8.4.

1.8.9 Corollary *If δ is a distance on X, then the family of pq-metrics*

$$\mathcal{H} \doteqdot \left\{ d_Z^\zeta \mid \zeta < \infty,\ Z \in 2^X \right\}$$

is a basis for the gauge associated with δ.

Proof This follows at once from the foregoing result and from the proof of 1.8.6 where it was implicitly shown that $\delta(x, A) = \sup_{d\in\mathcal{H}} \inf_{a\in A} d(x,a)$.

□

Next we study the relationship between gauges and approach systems.

1.8.10 Theorem (G $\rightrightarrows$ A) *If $\mathcal{H} \subset pq\mathrm{M}^\infty(X)$ is a gauge basis on X, then $(\mathcal{B}(x))_{x\in X}$, where, for each $x \in X$,*

$$\mathcal{B}(x) \doteqdot \{d(x,\cdot) \mid d \in \mathcal{H}\},$$

is an approach basis on X. Moreover, if $\mathcal{G}$ denotes the gauge generated by $\mathcal{H}$ and $(\mathcal{A}(x))_{x\in X}$ denotes the approach system generated by $(\mathcal{B}(x))_{x\in X}$, then

$$\mathcal{G} = \{d \in pq\mathrm{M}^\infty(X) \mid \forall x \in X : d(x,\cdot) \in \mathcal{A}(x)\}.$$

Proof That, for each $x \in X$, $\mathcal{B}(x)$ is an ideal basis follows from the fact that $\mathcal{G}$ is a basis for an ideal in $pq\mathrm{M}^{\infty}(X)$, and that it satisfies (B1) and (B2) follows from the fact that the members of $\mathcal{G}$ are ∞pq-metrics. To prove the final claim of the theorem it suffices to notice that

$$\begin{aligned} d \in \mathcal{G} &\Leftrightarrow \forall x \in X,\ \forall \epsilon > 0,\ \forall \omega < \infty\ \exists d' \in \mathcal{H} : d(x,\cdot) \wedge \omega \leq d'(x,\cdot) + \epsilon \\ &\Leftrightarrow \forall x \in X,\ \forall \epsilon > 0,\ \forall \omega < \infty\ \exists \varphi \in \mathcal{B}(x) : d(x,\cdot) \wedge \omega \leq \varphi + \epsilon \\ &\Leftrightarrow \forall x \in X : d(x,\cdot) \in \mathcal{A}(x). \end{aligned}$$

□

1.8.11 Theorem (A $\rightrightarrows$ G) *If $(\mathcal{A}(x))_{x \in X}$ is an approach system on X, then*

$$\mathcal{G} \doteqdot \{d \in pq\mathrm{M}^{\infty}(X) \mid \forall x \in X : d(x,\cdot) \in \mathcal{A}(x)\}$$

is a gauge on X.

Proof That $\mathcal{G}$ is an ideal in $pq\mathrm{M}^{\infty}(X)$ follows from the fact that, for each $x \in X$, $\mathcal{A}(x)$ is an ideal. That $\mathcal{G}$ is saturated follows from the fact that, for each $x \in X$, $\mathcal{A}(x)$ is saturated.

□

In order to conclude our proof that gauges and approach systems are equivalent concepts we need some supplementary results. In particular, to be able to work with the gauge associated with an approach system, as given by the foregoing theorem, we again require a specified set of ∞pq-metrics in that gauge. The following proposition provides us with such a collection.

1.8.12 Proposition *If $(\mathcal{A}(x))_{x \in X}$ is an approach basis on X, then, for any $\zeta < \infty$ and $Z \subset X$, the function*

$$d_Z^{\zeta} : X \times X \to [0,\infty] : (x,y) \to \left(\sup_{\varphi \in \mathcal{A}(x)} \inf_{z \in Z} \varphi(z) \wedge \zeta - \sup_{\psi \in \mathcal{A}(y)} \inf_{z \in Z} \psi(z) \wedge \zeta \right) \vee 0$$

is a pq-metric in the gauge associated with $\mathcal{A}$.

Proof That d_Z^{ζ} is a pq-metric is easily seen. Let $x \in X$ and $\epsilon > 0$ be fixed. Choose $\varphi_0 \in \mathcal{A}(x)$ such that

$$\sup_{\varphi \in \mathcal{A}(x)} \inf_{z \in Z} \varphi(z) \wedge \zeta \leq \inf_{z \in Z} \varphi_0(z) \wedge \zeta + \epsilon$$

and then choose a family $(\psi_u)_{u\in X} \in \prod_{u\in X} \mathcal{A}(u)$ such that, for all $u, z \in X$,

$$\varphi_0(z) \wedge \zeta \leq \psi_x(u) + \psi_u(z) + \epsilon.$$

Then it follows that, for all $y \in X$,

$$\begin{aligned}
d_Z^\zeta(x,y) &= \left(\sup_{\varphi\in\mathcal{A}(x)} \inf_{z\in Z} \varphi(z) \wedge \zeta - \sup_{\psi\in\mathcal{A}(y)} \inf_{z\in Z} \psi(z) \wedge \zeta\right) \vee 0 \\
&\leq \left(\inf_{z\in Z} \varphi_0(z) \wedge \zeta + \epsilon - \inf_{z\in Z} \psi_y(z) \wedge \zeta\right) \vee 0 \\
&\leq \left(\inf_{z\in Z} (\psi_x(y) + \psi_y(z) + \epsilon) \wedge \zeta + \epsilon - \inf_{z\in Z} \psi_y(z) \wedge \zeta\right) \vee 0 \\
&\leq \left(\inf_{z\in Z} (\psi_x(y) + \psi_y(z) \wedge \zeta + \epsilon) + \epsilon - \inf_{z\in Z} \psi_y(z) \wedge \zeta\right) \vee 0 \\
&\leq \left(\psi_x(y) + 2\epsilon + \inf_{z\in Z} \psi_y(z) \wedge \zeta - \inf_{z\in Z} \psi_y(z) \wedge \zeta\right) \vee 0 \\
&= \psi_x(y) + 2\epsilon,
\end{aligned}$$

which by (A2) proves that, for all $x \in X$, $d_Z^\zeta(x,\cdot) \in \mathcal{A}(x)$. Hence d_Z^ζ is indeed a member of the gauge associated with $\mathcal{A}$.

□

The similarity between the *pq*-metrics of the foregoing proposition and those of 1.8.4 is not just a coincidence. Actually it will follow from 1.8.24 that they are the same. Hence we are justified in using the same notation.

1.8.13 Proposition *If $(\mathcal{A}(x))_{x\in X}$ is an approach system on X and $\mathcal{G}$ is the associated gauge, then we have*

$$\sup_{\varphi\in\mathcal{A}(x)} \inf_{a\in A} \varphi(a) = \sup_{d\in\mathcal{G}} \inf_{a\in A} d(x,a) \qquad \forall x \in X,\ \forall A \subset X.$$

Proof One inequality is an immediate consequence of the definition of $\mathcal{G}$. Making use of 1.8.12 we further obtain

$$\begin{aligned}
\sup_{d\in\mathcal{G}}\inf_{a\in A} d(x,a) &\geq \sup_{\zeta<\infty}\sup_{Z\subset X}\inf_{a\in A} d_Z^{\zeta}(x,a)\\
&= \sup_{\zeta<\infty}\sup_{Z\subset X}\inf_{a\in A}\left(\sup_{\varphi\in\mathcal{A}(x)}\inf_{z\in Z}\varphi(z)\wedge\zeta - \sup_{\psi\in\mathcal{A}(a)}\inf_{z\in Z}\psi(z)\wedge\zeta\right)\vee 0\\
&\geq \sup_{\zeta<\infty}\inf_{a\in A}\left(\sup_{\varphi\in\mathcal{A}(x)}\inf_{z\in A}\varphi(z)\wedge\zeta - \sup_{\psi\in\mathcal{A}(a)}\inf_{z\in A}\psi(z)\wedge\zeta\right)\vee 0\\
&= \sup_{\zeta<\infty}\inf_{a\in A}\sup_{\varphi\in\mathcal{A}(x)}\inf_{z\in A}\varphi(z)\wedge\zeta\\
&= \sup_{\varphi\in\mathcal{A}(x)}\inf_{z\in A}\varphi(z),
\end{aligned}$$

which proves the other inequality.

□

1.8.14 Theorem *If $(\mathcal{A}(x))_{x\in X}$ is an approach system on X and $\mathcal{G}$ is the associated gauge, then, for all $x \in X$, we have that $\mathcal{A}(x) = \widehat{\mathcal{B}(x)}$, where*

$$\mathcal{B}(x) = \{d(x,\cdot) \mid d \in \mathcal{G}\}.$$

Proof It is immediately clear that, for all $x \in X$, $\widehat{\mathcal{B}(x)} \subset \mathcal{A}(x)$. To show that the converse inclusion also holds we make use of 1.8.13. Suppose that, for some $x \in X$, there exists $\psi \in \mathcal{A}(x) \setminus \widehat{\mathcal{B}(x)}$. Hence ψ is not dominated by $\mathcal{B}(x)$, which implies that we can find $\epsilon > 0$ and $\omega < \infty$ such that, for all $d \in \mathcal{G}$,

$$\psi \wedge \omega \not\leq d(x,\cdot) + \epsilon.$$

For each $d \in \mathcal{G}$, let $A(d) \doteqdot \{y \in X \mid \psi(y) \wedge \omega > d(x,y) + \epsilon\}$. Then it is clear that, for all $d, e \in \mathcal{G}$, $A(d) \cap A(e) = A(d \vee e) \neq \emptyset$. Hence we obtain

$$\begin{aligned}
\sup_{d\in\mathcal{G}}\sup_{e\in\mathcal{G}}\inf_{y\in A(d)} e(x,y)\wedge\omega &\leq \sup_{d\in\mathcal{G}}\sup_{e\in\mathcal{G}}\inf_{y\in A(d\vee e)} (d\vee e)(x,y)\wedge\omega\\
&= \sup_{d\in\mathcal{G}}\inf_{y\in A(d)} d(x,y)\wedge\omega\\
&\leq \sup_{d\in\mathcal{G}}\inf_{y\in A(d)} (\psi(y)\wedge\omega - \epsilon)\wedge\omega\\
&\leq \sup_{d\in\mathcal{G}}\inf_{y\in A(d)} \psi(y)\wedge\omega - \epsilon\\
&\leq \sup_{d\in\mathcal{G}}\sup_{\varphi\in\mathcal{A}(x)}\inf_{y\in A(d)} \varphi(y)\wedge\omega - \epsilon,
\end{aligned}$$

which, by 1.8.13, is a contradiction.

□

The combined results of 1.8.10, 1.8.11, and 1.8.14 show that gauges and approach systems are equivalent systems.

Now we come to the relationship between distances and hulls. For the transition from a distance to a hull we go via the gauge associated with the distance. Therefore we first give the formula for the transition from gauges to hulls.

1.8.15 Theorem (G $\rightrightarrows$ H) *If $\mathcal{G} \subset pq\mathrm{M}^{\infty}(X)$ is a gauge basis on X, then the function $h : [0,\infty]^X \longrightarrow [0,\infty]^X$, defined by*

$$h(\mu)(x) \doteqdot \sup_{d\in\mathcal{G}} \inf_{y\in X} (\mu(y) + d(x,y)) \qquad \forall \mu \in [0,\infty]^X,\ \forall x \in X,$$

is a hull operator on X.

Proof That h fulfills (H1), (H2), and (H4) is immediately seen. To prove (H3), let $\mu \in [0,\infty]^X$. That $h(h(\mu)) \leq h(\mu)$ follows from (H1). To show the converse inequality let $x \in X$; then it follows that

$$\begin{aligned} h(h(\mu))(x) &= \sup_{d\in\mathcal{G}} \inf_{y\in X} (h(\mu)(y) + d(x,y)) \\ &= \sup_{d\in\mathcal{G}} \inf_{y\in X} \left(\sup_{d'\in\mathcal{G}} \inf_{z\in X} (\mu(z) + d'(y,z)) + d(x,y) \right) \\ &\geq \sup_{d\in\mathcal{G}} \inf_{y\in X} \left(\inf_{z\in X} (\mu(z) + d(y,z)) + d(x,y) \right) \\ &\geq \sup_{d\in\mathcal{G}} \inf_{z\in X} (\mu(z) + d(x,z)) \\ &= h(\mu)(x). \end{aligned}$$

□

1.8.16 Theorem (D $\rightrightarrows$ H) *If $\delta : X \times 2^X \longrightarrow [0,\infty]$ is a distance on X, $\mathcal{G}$ is the gauge associated with δ, and h is the hull associated with $\mathcal{G}$, then we have that*

$$\delta(x,A) = h(\theta_A)(x) \qquad \forall x \in X,\ \forall A \subset X.$$

Proof This is an immediate consequence of 1.8.6 and 1.8.15.

□

1.8.17 Theorem (H $\rightrightarrows$ D) *If $h : [0,\infty]^X \longrightarrow [0,\infty]^X$ is a hull operator on X, then the function*

$$\delta : X \times 2^X \longrightarrow [0,\infty] : (x, A) \longrightarrow h(\theta_A)(x)$$

is a distance on X.

Proof Properties (D1), (D2), and (D3) are immediate. To prove (D4) first remark that, for any $A \subset X$ and $\epsilon \in \mathbb{R}^+$, it follows from the definition of $A^{(\epsilon)}$ that

$$h(\theta_A) \leq \theta_{A^{(\epsilon)}} + \epsilon,$$

and consequently, it follows that

$$\begin{aligned}\delta_A &= h(\theta_A)\\ &= h(h(\theta_A))\\ &\leq h(\theta_{A^{(\epsilon)}} + \epsilon)\\ &= h(\theta_{A^{(\epsilon)}}) + \epsilon\\ &= \delta_{A^{(\epsilon)}} + \epsilon.\end{aligned}$$

□

1.8.18 Theorem *If $h : [0,\infty]^X \longrightarrow [0,\infty]^X$ is a hull operator on X, δ is the distance associated with h, and $\mathcal{G}$ is the gauge associated with δ, then*

$$h(\mu)(x) = \sup_{d\in\mathcal{G}} \inf_{y\in X} (\mu(y) + d(x,y)) \qquad \forall \mu \in [0,\infty]^X,\ \forall x \in X.$$

Proof If we define the function $h' : [0,\infty]^X \longrightarrow [0,\infty]^X$ by

$$h'(\mu)(x) = \sup_{d\in\mathcal{G}} \inf_{y\in X} (\mu(y) + d(x,y)) \qquad \forall \mu \in [0,\infty]^X,\ \forall x \in X,$$

then it follows from 1.8.15 that h' is a hull operator. By 1.6.5 we know that hull operators are completely determined by their restriction to $Ind(X)$. Hence, to prove the theorem, it suffices to notice that by 1.8.6, for any $x \in X$ and $A \subset X$,

$$\begin{aligned}h'(\theta_A)(x) &= \sup_{d\in\mathcal{G}} \inf_{y\in X} (\theta_A(y) + d(x,y))\\ &= \sup_{d\in\mathcal{G}} \inf_{y\in A} d(x,y)\\ &= \delta(x, A)\\ &= h(\theta_A)(x).\end{aligned}$$

□

The combined results of 1.8.15, 1.8.16, 1.8.17, and 1.8.18 show that distances and hull operators are equivalent systems.

We now turn our attention to the relationship between distances and towers.

1.8.19 Theorem (D $\rightrightarrows$ T) *If $\delta : X \times 2^X \longrightarrow [0,\infty]$ is a distance on X, then the family $(\mathrm{t}_\epsilon)_{\epsilon\in\mathbb{R}^+}$, defined by*

$$\mathrm{t}_\epsilon(A) \doteqdot A^{(\epsilon)} \qquad \forall A \subset X, \forall \epsilon \in \mathbb{R}^+,$$

is a tower on X. Moreover, for any $x \in X$ and $A \subset X$, we have

$$\delta(x, A) = \inf\{\epsilon \in \mathbb{R}^+ \mid x \in \mathrm{t}_\epsilon(A)\}.$$

Proof (T1), (T2), and (T3) follow easily from (D1), (D2), and (D3). To prove (T4) let $A \subset X$, $\epsilon, \gamma \in \mathbb{R}^+$, and $x \in \mathrm{t}_\epsilon(\mathrm{t}_\gamma(A))$. This implies that $\delta\left(x, A^{(\gamma)}\right) \leq \epsilon$ and thus, by (D4), that $\delta(x, A) \leq \epsilon + \gamma$, i.e. $x \in \mathrm{t}_{\epsilon+\gamma}(A)$. (T5) finally follows from the fact that $\delta(x, A) \leq \epsilon$ if and only if, for all $\epsilon' > \epsilon$, $\delta(x, A) \leq \epsilon'$. The last claim of the theorem follows at once from the definitions.

□

1.8.20 Theorem (T $\rightrightarrows$ D) *If $(\mathrm{t}_\epsilon)_{\epsilon\in\mathbb{R}^+}$ is a tower on X, then the function*

$$\delta : X \times 2^X \longrightarrow [0, \infty],$$

defined by

$$\delta(x, A) \doteqdot \inf\{\epsilon \in \mathbb{R}^+ \mid x \in \mathrm{t}_\epsilon(A)\} \qquad \forall x \in X, \forall A \subset X,$$

is a distance on X. Moreover, we have that

$$\mathrm{t}_\epsilon(A) = A^{(\epsilon)} \qquad \forall \epsilon \in \mathbb{R}^+, \forall A \subset X.$$

Proof (D1), (D2), and (D3) follow easily from (T1), (T2), and (T3). To prove (D4) we first prove the last claim of the theorem. Let $x \in X$, $A \subset X$, and $\epsilon \in \mathbb{R}^+$; then we have

$$\begin{aligned} x \in A^{(\epsilon)} &\Rightarrow \inf\{\alpha \in \mathbb{R}^+ \mid x \in \mathrm{t}_\alpha(A)\} \leq \epsilon \\ &\Rightarrow \forall \alpha > \epsilon : \ x \in \mathrm{t}_\alpha(A) \\ &\Rightarrow x \in \bigcap_{\alpha > \epsilon} \mathrm{t}_\alpha(A) = \mathrm{t}_\epsilon(A). \end{aligned}$$

Conversely we have

$$x \in \mathfrak{t}_\epsilon(A) \Rightarrow \delta(x, A) = \inf\left\{\alpha \in \mathbb{R}^+ \mid x \in \mathfrak{t}_\alpha(A)\right\} \leq \epsilon$$
$$\Rightarrow x \in A^{(\epsilon)}.$$

(D4) now follows from the observation that if $\delta\left(x, A^{(\epsilon)}\right) < \alpha$ then $x \in \mathfrak{t}_\alpha\left(A^{(\epsilon)}\right) = \mathfrak{t}_\alpha(\mathfrak{t}_\epsilon(A)) \subset \mathfrak{t}_{\alpha+\epsilon}(A)$ and thus $\delta(x, A) \leq \alpha + \epsilon$.

□

Finally we turn to the relationship between regular function frames and hull operators.

1.8.21 Theorem (R $\rightrightarrows$ H) *If $\mathcal{R}$ is a regular function frame on X, then the function $h : [0,\infty]^X \longrightarrow [0,\infty]^X$, defined by*

$$h(\mu) \doteqdot \sup\{\nu \in \mathcal{R} \mid \nu \leq \mu\} \qquad \forall \mu \in [0,\infty]^X,$$

is a hull operator on X. Moreover,

$$\mathcal{R} = \left\{\mu \in [0,\infty]^X \mid h(\mu) = \mu\right\}.$$

Proof This is perfectly similar to the analogous result in topological spaces concerning the relationship between closed sets and closure operator, with the exception of (H4). Let $\mu \in [0,\infty]^X$ and let $\alpha \in [0,\infty]$; then we have

$$\begin{aligned} h(\mu+\alpha) &= \sup\{\xi \in \mathcal{R} \mid \xi \leq \mu+\alpha\} \\ &= \sup\{\xi \in \mathcal{R} \mid \alpha \leq \xi \leq \mu+\alpha\} \\ &= \sup\{\xi-\alpha \mid \xi \in \mathcal{R},\ \alpha \leq \xi,\ \xi-\alpha \leq \mu\} + \alpha \\ &= \sup\{\rho \in \mathcal{R} \mid \rho \leq \mu\} + \alpha \\ &= h(\mu) + \alpha. \end{aligned}$$

□

1.8.22 Theorem (H $\rightrightarrows$ R) *If $h : [0,\infty]^X \longrightarrow [0,\infty]^X$ is a hull operator on X, then*

$$\mathcal{R} \doteqdot \left\{\mu \in [0,\infty]^X \mid h(\mu) = \mu\right\}$$

is a regular function frame on X. Moreover, we have that

$$h(\mu) = \sup\{\nu \in \mathcal{R} \mid \nu \leq \mu\} \qquad \forall \mu \in [0,\infty]^X.$$

Proof Again, this is perfectly similar to the analogous result in topological spaces concerning the relationship between closed sets and closure operator, with the exception of (R3) and (R4). Let $\mu \in \mathcal{R}$. If $\alpha \in [0,\infty]$, then we have

$$\begin{aligned} h(\mu+\alpha) &= h(\mu)+\alpha \\ &= \mu+\alpha, \end{aligned}$$

and hence $\mu+\alpha \in \mathcal{R}$. If $\alpha \in \left[0, \inf_{x\in X} \mu(x)\right]$, then it follows from 1.6.2 that $h(\mu-\alpha) = \mu-\alpha$ and hence again $\mu-\alpha \in \mathcal{R}$.

□

We have now seen all the basic transitions which prove that the seven different types of structures are equivalent with each other. For the transition from distances to hull operators we went via gauges. Often, however, it will be necessary to have a noncircuitous transition from one structure to another. In the rest of this section we will derive some such noncircuitous transitions which we will require in the sequel.

1.8.23 Theorem (D $\rightrightarrows$ A) *If $\delta : X \times 2^X \longrightarrow [0,\infty]$ is a distance on X, then the associated approach system $(\mathcal{A}(x))_{x\in X}$ is given by*

$$\mathcal{A}(x) = \left\{\varphi \in [0,\infty]^X \mid \forall A \subset X : \inf_{a\in A} \varphi(a) \leq \delta(x,A)\right\} \qquad \forall x \in X.$$

Proof It follows from 1.8.3 and 1.8.10 that the approach system $(\mathcal{A}(x))_{x\in X}$ associated with δ is generated by the basis

$$\mathcal{B}(x) = \left\{d(x,\cdot) \mid d \in pq\mathbf{M}^\infty(X), \forall A \subset X : \inf_{a\in A} d(x,a) \leq \delta(x,A)\right\} \qquad \forall x \in X.$$

For all $x \in X$, let us put

$$\mathcal{C}(x) \doteqdot \left\{\varphi \in [0,\infty]^X \mid \forall A \subset X : \inf_{a\in A} \varphi(a) \leq \delta(x,A)\right\}.$$

Then it is immediately clear that, for all $x \in X$, $\mathcal{A}(x) \subset \mathcal{C}(x)$. Let $\mathcal{G}$ stand for the gauge associated with δ and suppose that there exists $\psi \in \mathcal{C}(x) \setminus \mathcal{A}(x)$; then it follows that ψ is not dominated by $\mathcal{B}(x)$ and hence there exist $\epsilon > 0$ and $\omega < \infty$ such that, for all $d \in \mathcal{G}$,

$$\psi \wedge \omega \not\leq d(x, \cdot) + \epsilon.$$

For each $d \in \mathcal{G}$, let $A(d) \doteqdot \{y \in X \mid \psi(y) \wedge \omega > d(x, y) + \epsilon\}$. Then it is clear that, for all $d, e \in \mathcal{G}$, $A(d) \cap A(e) = A(d \vee e) \neq \emptyset$. Hence, from 1.8.6, we obtain

$$\begin{aligned}
\sup_{d\in\mathcal{G}} \delta(x, A(d)) \wedge \omega &= \sup_{d\in\mathcal{G}} \sup_{e\in\mathcal{G}} \inf_{y\in A(d)} e(x, y) \wedge \omega \\
&\leq \sup_{d\in\mathcal{G}} \sup_{e\in\mathcal{G}} \inf_{y\in A(d\vee e)} (d \vee e)(x, y) \wedge \omega \\
&= \sup_{d\in\mathcal{G}} \inf_{y\in A(d)} d(x, y) \wedge \omega \\
&\leq \sup_{d\in\mathcal{G}} \inf_{y\in A(d)} (\psi(y) \wedge \omega - \epsilon) \wedge \omega \\
&\leq \sup_{d\in\mathcal{G}} \inf_{y\in A(d)} \psi(y) \wedge \omega - \epsilon \\
&\leq \sup_{d\in\mathcal{G}} \sup_{\varphi\in\mathcal{A}(x)} \inf_{y\in A(d)} \varphi(y) \wedge \omega - \epsilon \\
&\leq \sup_{d\in\mathcal{G}} \delta(x, A(d)) \wedge \omega - \epsilon,
\end{aligned}$$

which is a contradiction.

□

1.8.24 Theorem (A ⇉ D) *If $(\mathcal{A}(x))_{x\in X}$ is an approach system on X, then the associated distance is given by*

$$\delta(x, A) = \sup_{\varphi\in\mathcal{A}(x)} \inf_{a\in A} \varphi(a) \qquad \forall x \in X,\ \forall A \subset X.$$

Proof This follows at once from 1.8.5 and 1.8.13.

□

1.8.25 Theorem (H ⇉ A) *If $h : [0, \infty]^X \longrightarrow [0, \infty]^X$ is a hull operator on X, then the associated approach system $(\mathcal{A}(x))_{x\in X}$ is given by*

$$\mathcal{A}(x) = \left\{\varphi \in [0, \infty]^X \mid \forall \mu \in [0, \infty]^X : \inf_{y\in X} (\mu + \varphi)(y) \leq h(\mu)(x)\right\} \qquad \forall x \in X.$$

Proof This follows from 1.8.23 and the fact that, by 1.6.6, if φ fulfills the property $\inf_{a\in A} \varphi(a) \leq h(\theta_A)(x)$, for all $A \subset X$, then it also fulfills the property $\inf_{y\in X} (\mu + \varphi)(y) \leq h(\mu)(x)$, for all $\mu \in [0,\infty]^X$, as can easily be verified.

□

1.8.26 Proposition *If $(\mathcal{A}(x))_{x\in X}$ is an approach system on X with basis $(\mathcal{B}(x))_{x\in X}$, then the following formulas hold:*

1. *for any $x \in X$ and $A \subset X$,*

$$\sup_{\varphi\in\mathcal{A}(x)} \inf_{a\in A} \varphi(a) = \sup_{\varphi\in\mathcal{B}(x)} \inf_{a\in A} \varphi(a),$$

2. *for any $x \in X$ and $\mu \in [0,\infty]^X$,*

$$\sup_{\varphi\in\mathcal{A}(x)} \inf_{y\in X} (\mu + \varphi)(y) = \sup_{\varphi\in\mathcal{B}(x)} \inf_{y\in X} (\mu + \varphi)(y),$$

3. *for any $x \in X$ and $\mathcal{F} \in \mathbf{F}(X)$,*

$$\sup_{\varphi\in\mathcal{A}(x)} \inf_{F\in\mathcal{F}} \sup_{y\in F} \varphi(y) = \sup_{\varphi\in\mathcal{B}(x)} \inf_{F\in\mathcal{F}} \sup_{y\in F} \varphi(y),$$

4. *for any $x \in X$ and $\mathcal{F} \in \mathbf{F}(X)$,*

$$\sup_{\varphi\in\mathcal{A}(x)} \sup_{F\in\mathcal{F}} \inf_{y\in F} \varphi(y) = \sup_{\varphi\in\mathcal{B}(x)} \sup_{F\in\mathcal{F}} \inf_{y\in F} \varphi(y).$$

Proof We only give the proof of the first formula; the other ones are precisely the same. One inequality is trivial. To prove the other inequality let $\varphi \in \mathcal{A}(x)$ be fixed. Then it follows from the definition of a basis for an approach system that there exists a family $(\varphi_\epsilon^\omega)_{\epsilon>0,\omega<\infty}$ in $\mathcal{B}(x)$ which dominates φ. From this it follows that, for all $\omega < \infty$,

$$\begin{aligned} \inf_{a\in A} \varphi(a) \wedge \omega &\leq \inf_{\epsilon>0} \inf_{a\in A} \varphi_\epsilon^\omega(a) + \epsilon \\ &\leq \inf_{\epsilon>0} \sup_{\psi\in\mathcal{B}(x)} \inf_{a\in A} \psi(a) + \epsilon \\ &= \sup_{\psi\in\mathcal{B}(x)} \inf_{a\in A} \psi(a). \end{aligned}$$

□

A similar result holds for gauges and gauge bases.

1.8.27 Proposition *If $\mathcal{G}$ is a gauge on X with basis $\mathcal{H}$, then the following formulas hold:*

1. *for any $x \in X$ and $A \subset X$,*

$$\sup_{d\in\mathcal{G}} \inf_{a\in A} d(x,a) = \sup_{d\in\mathcal{H}} \inf_{a\in A} d(x,a),$$

2. *for any $x \in X$ and $\mu \in [0,\infty]^X$,*

$$\sup_{d\in\mathcal{G}} \inf_{y\in X} (\mu(y) + d(x,y)) = \sup_{d\in\mathcal{H}} \inf_{y\in X} (\mu(y) + d(x,y)),$$

3. *for any $x \in X$ and $\mathcal{F} \in \mathbf{F}(X)$,*

$$\sup_{d\in\mathcal{G}} \inf_{F\in\mathcal{F}} \sup_{y\in F} d(x,y) = \sup_{d\in\mathcal{H}} \inf_{F\in\mathcal{F}} \sup_{y\in F} d(x,y),$$

4. *for any $x \in X$ and $\mathcal{F} \in \mathbf{F}(X)$,*

$$\sup_{d\in\mathcal{G}} \sup_{F\in\mathcal{F}} \inf_{y\in F} d(x,y) = \sup_{d\in\mathcal{H}} \sup_{F\in\mathcal{F}} \inf_{y\in F} d(x,y).$$

Proof This is precisely the same as the proof of 1.8.26.

□

1.8.28 Theorem (A $\rightrightarrows$ H) *If $(\mathcal{A}(x))_{x\in X}$ is an approach system on X, then the associated hull operator is given by*

$$h(\mu)(x) = \sup_{\varphi\in\mathcal{A}(x)} \inf_{y\in X} (\mu + \varphi)(y) \qquad \forall x \in X,\ \forall \mu \in [0,\infty]^X.$$

Proof This follows at once from 1.8.14, 1.8.15, and 1.8.26.

□

Before we can describe the transition from approach systems to limit operators, we need a supplementary result.

1.8.29 Proposition *If $\mathcal{U}$ is an ultrafilter on X and $f : X \longrightarrow [0,\infty]$ is an arbitrary function, then*

$$\sup_{U\in\mathcal{U}} \inf_{y\in U} f(y) = \inf_{U\in\mathcal{U}} \sup_{y\in U} f(y).$$

Proof It is immediately clear that, for any $U, V \in \mathcal{U}$, we have $\inf_{y \in U} f(y) \leq \sup_{y \in V} f(y)$. Consequently

$$\sup_{U \in \mathcal{U}} \inf_{y \in U} f(y) \leq \inf_{U \in \mathcal{U}} \sup_{y \in U} f(y).$$

To prove the other inequality, suppose that $\inf_{U \in \mathcal{U}} \sup_{y \in U} f(y) > r$. Then, for all $U \in \mathcal{U}$, we have $U \cap \{f > r\} \neq \emptyset$ and consequently $\{f > r\} \in \mathcal{U}$. This then also implies that $r \leq \inf_{y \in \{f > r\}} f(y)$, from which the converse inequality follows.

□

1.8.30 Theorem (A ⇉ L) *If $(\mathcal{A}(x))_{x \in X}$ is an approach system on X, then the associated limit operator is given by*

$$\lambda \mathcal{F}(x) = \sup_{\varphi \in \mathcal{A}(x)} \inf_{F \in \mathcal{F}} \sup_{y \in F} \varphi(y) \qquad \forall x \in X, \forall \mathcal{F} \in \mathbf{F}(X).$$

Proof Let δ be the distance associated with $(\mathcal{A}(x))_{x \in X}$. Applying complete distributivity, 1.8.1, 1.8.24, and 1.8.29 we obtain

$$\begin{aligned}
\lambda \mathcal{F}(x) &= \sup_{\mathcal{U} \in \mathbf{U}(\mathcal{F})} \sup_{U \in \mathcal{U}} \delta(x, U) \\
&= \sup_{\mathcal{U} \in \mathbf{U}(\mathcal{F})} \sup_{U \in \mathcal{U}} \sup_{\varphi \in \mathcal{A}(x)} \inf_{y \in U} \varphi(y) \\
&= \sup_{\mathcal{U} \in \mathbf{U}(\mathcal{F})} \sup_{\varphi \in \mathcal{A}(x)} \inf_{U \in \mathcal{U}} \sup_{y \in U} \varphi(y) \\
&= \sup_{\varphi \in \mathcal{A}(x)} \inf_{(A_{\mathcal{U}})_{\mathcal{U} \in \mathbf{U}(\mathcal{F})} \in \prod_{\mathcal{U} \in \mathbf{U}(\mathcal{F})} \mathcal{U}} \sup_{\mathcal{U} \in \mathbf{U}(\mathcal{F})} \sup_{y \in A_{\mathcal{U}}} \varphi(y) \\
&= \sup_{\varphi \in \mathcal{A}(x)} \inf_{(A_{\mathcal{U}})_{\mathcal{U} \in \mathbf{U}(\mathcal{F})} \in \prod_{\mathcal{U} \in \mathbf{U}(\mathcal{F})} \mathcal{U}} \sup_{y \in \bigcup_{\mathcal{U} \in \mathbf{U}(\mathcal{F})} A_{\mathcal{U}}} \varphi(y) \\
&= \sup_{\varphi \in \mathcal{A}(x)} \inf_{F \in \mathcal{F}} \sup_{y \in F} \varphi(y).
\end{aligned}$$

□

1.8.31 Theorem (G ⇉ L) *If $\mathcal{G} \subset pq\mathrm{M}^{\infty}(X)$ is a gauge on X, then the associated limit operator is given by*

$$\lambda \mathcal{F}(x) = \sup_{d \in \mathcal{G}} \inf_{F \in \mathcal{F}} \sup_{y \in F} d(x, y) \qquad \forall x \in X, \forall \mathcal{F} \in \mathbf{F}(X).$$

Proof This follows at once from 1.8.14, 1.8.30, and 1.8.26.

□

We have now seen all the main transitions for going from one basic structure to another which we will require in this work. Clearly there are a number of fundamental transitions which have great conceptual interest and technical elegance, whereas some are no more than the composition of other transitions. Again this is not unlike the situation in topology, where defining, for example, the open sets from the neighborhoods goes in a direct way, whereas defining them in terms of the closure operator goes most naturally via the closed sets.

1.8.32 Definition A pair $(X, \mathfrak{S})$, where $\mathfrak{S}$ is a distance, a limit operator, an approach system, a gauge, a tower, a hull operator, or a regular function frame, is called an *approach space*.

As we have just seen, an approach space can be determined by giving any of these equivalent structures and in most of the examples which we will study later on, at least one of them will present itself as the basic one from which the others can be derived. If no confusion is possible, we will always denote a distance by δ and the associated limit operator, approach system, gauge, tower, hull operator, and regular function frame respectively by λ, $(\mathcal{A}(x))_{x\in X}$, $\mathcal{G}$, $\mathfrak{t} \doteqdot (\mathfrak{t}_\epsilon)_{\epsilon\in\mathbb{R}^+}$, h, and $\mathcal{R}$. It will generally be understood that these structures are associated with one another. If we are dealing with different approach spaces at the same time, we will make clear which structures are associated with each other, by using some appropriate indices or accents.

Sometimes we will also denote an approach space simply by its underlying set. Some spaces may get special notational treatment like those in the next example. When an approach space is determined by a gauge basis (or by an approach basis), then we will usually refer to the other structures of the approach space as being *generated* by that base.

We use both notations $A^{(\epsilon)}$ and $\mathfrak{t}_\epsilon(A)$ for $\{x \in X \mid \delta(x, A) \leq \epsilon\}$. In all that follows we will freely use the different structures defining approach spaces as well as the transitions among them. At the end of this section, for easy reference, we give some tables containing the most important transition formulas.

Once we have developed more machinery, we will encounter many individual examples, as well as special classes of approach spaces. Two special examples which we can easily introduce at this stage are the following. They both have $[0, \infty]$ as the underlying set.

It will be useful to have at our disposal the following ∞pq-metrics. We denote by $d_{\mathbb{E}}$ the ∞-metric defined by

$$d_{\mathbb{E}} : [0, \infty] \times [0, \infty] \longrightarrow [0, \infty] : (x, y) \longrightarrow |x - y|$$

and by $d_{\mathbb{P}}$ the ∞q-metric defined by

$$d_{\mathbb{P}} : [0, \infty] \times [0, \infty] \longrightarrow [0, \infty] : (x, y) \longrightarrow (x - y) \vee 0.$$

Note that both are straightforward extensions of well-known ∞pq-metrics on $[0, \infty[$. The restriction of $d_{\mathbb{E}}$ to $[0, \infty[$ will be denoted by d_E (this is the usual Euclidean metric) and the restriction of $d_{\mathbb{P}}$ to $[0, \infty[$ will be denoted by d_P.

For the topologies generated by $d_{\mathbb{E}}$ and by $d_{\mathbb{P}}$ the point ∞ is isolated.

1.8.33 Examples 1. Define

$$\delta_{\mathbb{E}} : [0, \infty] \times 2^{[0,\infty]} \longrightarrow [0, \infty]$$

by

$$\delta_{\mathbb{E}}(x, A) \doteqdot \begin{cases} 0 & x = \infty, A \text{ unbounded}, \\ \infty & x = \infty, A \text{ bounded}, \\ \inf\limits_{a \in A} |x - a| & x < \infty. \end{cases}$$

It is easily verified that this function is a distance on $[0, \infty]$ and we will denote the approach space thus obtained by

$$\mathbb{E} \doteqdot ([0, \infty], \delta_{\mathbb{E}}).$$

This distance actually extends the Euclidean metric d_E to the Alexandroff compactification of $[0, \infty[$. The space $[0, \infty]$ equipped with the topology of the Alexandroff compactification is of course metrizable, but since the Euclidean metric on $[0, \infty[$ is unbounded, this topology can of course not be metrized by a metric which extends the Euclidean metric. This, however, is precisely what the distance $\delta_{\mathbb{E}}$ does. For $x \in [0, \infty[$ and $A \subset [0, \infty[$ we have that $\delta_{\mathbb{E}}(x, A) = \inf\limits_{x \in A} d_E(x, a)$, and so, on $[0, \infty[$, $\delta_{\mathbb{E}}$ is completely determined by d_E. For a more formal treatment of distances derived from a metric in this way we refer to the third chapter.

The approach system associated with $\delta_{\mathbb{E}}$ is the following:

$$\mathcal{A}_{\mathbb{E}}(x) \doteq \left\{\varphi \in [0,\infty]^{[0,\infty]} \mid \varphi \leq d_{\mathbb{E}}(x,\cdot)\right\} \qquad \forall x < \infty$$

and

$$\mathcal{A}_{\mathbb{E}}(\infty) \doteq \widehat{\mathcal{B}},$$

where

$$\mathcal{B} \doteq \left\{\theta_{]a,\infty]} \mid a < \infty\right\}.$$

So we see that in finite points the structure is determined, essentially, by the Euclidean metric and in the point ∞ the structure is determined by the neighborhoods of ∞ in the Alexandroff compactification.

We leave it to the reader to determine the other structures associated with this distance.

2. An analogous example, which we will use several times in the sequel, is obtained as follows. Define

$$\delta_{\mathbb{P}} : [0,\infty] \times 2^{[0,\infty]} \longrightarrow [0,\infty]$$

by

$$\delta_{\mathbb{P}}(x, A) \doteq \begin{cases} (x - \sup A) \vee 0 & A \neq \emptyset, \\ \infty & A = \emptyset. \end{cases}$$

Again this function is a distance on $[0,\infty]$ and we will denote the approach space thus defined by

$$\mathbb{P} \doteq ([0,\infty], \delta_{\mathbb{P}}).$$

The approach system associated with $\delta_{\mathbb{P}}$ is the following:

$$\mathcal{A}_{\mathbb{P}}(x) \doteq \left\{\varphi \in [0,\infty]^{[0,\infty]} \mid \varphi \leq d_{\mathbb{P}}(x,\cdot)\right\} \qquad \forall x < \infty$$

and

$$\mathcal{A}_{\mathbb{P}}(\infty) \doteq \widehat{\mathcal{B}},$$

where, again,

$$\mathcal{B} \doteq \left\{\theta_{]a,\infty]} \mid a < \infty\right\}.$$

The limit operator associated with this distance is determined as follows. Given any filter $\mathcal{F}$ on $[0, \infty]$, let

$$l(\mathcal{F}) \doteq \inf_{U \in \sec \mathcal{F}} \sup U.$$

Then

$$\lambda_{\mathbb{P}}\mathcal{F}(x) = (x - l(\mathcal{F})) \vee 0 \qquad \forall x \in [0, \infty].$$

Notice for example that for any filter $\mathcal{F}$, either $\lambda_{\mathbb{P}}\mathcal{F}(\infty) = 0$ or $\lambda_{\mathbb{P}}\mathcal{F}(\infty) = \infty$. The first case will occur if $l(\mathcal{F}) = \infty$, i.e. if every member of $\sec\mathcal{F}$ is unbounded.

Again we will leave it as an exercise for the reader to determine the other structures associated with this distance.

There is still another structure which can be associated with an approach space and which could actually also be axiomatized in order to give an equivalent defining structure. However, we will not be interested in doing this here. We do nevertheless need to be able to use this structure in what follows and hence we now define it in terms of different approach structures.

1.8.34 Definition Let (X, δ) be an approach space. We define what we will call the *adherence operator* (associated with δ and with all of the other defining structures) as the function

$$\alpha\colon \mathbf{F}(X) \longrightarrow [0, \infty]^X,$$

determined by

$$\alpha\mathcal{F}(x) \doteq \sup_{F \in \mathcal{F}} \delta(x, F) \qquad \forall x \in X,\ \forall \mathcal{F} \in \mathbf{F}(X).$$

The function

$$\alpha\mathcal{F} : X \longrightarrow [0, \infty]$$

is called the *adherence of* $\mathcal{F}$.

The interpretation of this operator is analogous to that of the limit operator. In each point $x \in X$, the value $\alpha\mathcal{F}(x)$ indicates how far the point x is away from being an adherence point of the filter $\mathcal{F}$.

A first justification of these interpretations is given in the following results, where we discover some basic relations between the concepts of limit operator and adherence operator which are reminiscent of the relations which exist between the notions of adherence point and limit point of a filter. Further justifications will be given in the second and especially in the third chapter.

1.8.35 Proposition *Let (X, δ) be an approach space. Then, for all $\mathcal{F}, \mathcal{G} \in \mathbf{F}(X)$, we have*

$$\mathcal{F} \subset \mathcal{G} \;\Rightarrow\; \alpha\mathcal{F} \leq \alpha\mathcal{G} \leq \lambda\mathcal{G} \leq \lambda\mathcal{F}.$$

Proof This follows from 1.8.1 and the definition of the adherence operator given in 1.8.34.

□

Notice that in a topological space, finer filters converge to more and adhere to fewer points. Comparing this to the result of 1.8.35 gives a first argument justifying our interpretation of the values of the limit and adherence of a filter.

1.8.36 Proposition *Let (X, δ) be an approach space. Then, for all $\mathcal{U} \in \mathbf{U}(X)$, we have*

$$\lambda\mathcal{U} = \alpha\mathcal{U}.$$

Proof If $\mathcal{U}$ is an ultrafilter, then $\sec\mathcal{U} = \mathcal{U}$ and, hence, the result follows from 1.8.1 and the definition of adherence operator given in 1.8.34.

□

The foregoing result generalizes the well-known fact that for ultrafilters limit points and adherence points are the same.

1.8.37 Proposition *Let (X, δ) be an approach space. Then, for all $\mathcal{F} \in \mathbf{F}(X)$ and $x \in X$, we have*

$$1.\ \lambda\mathcal{F}(x) = \sup_{\mathcal{U}\in\mathbf{U}(\mathcal{F})} \lambda\mathcal{U}(x) = \sup_{\mathcal{U}\in\mathbf{U}(\mathcal{F})} \alpha\mathcal{U}(x)$$

and

$$2.\ \alpha\mathcal{F}(x) = \inf_{\mathcal{U}\in\mathbf{U}(\mathcal{F})} \lambda\mathcal{U}(x) = \inf_{\mathcal{U}\in\mathbf{U}(\mathcal{F})} \alpha\mathcal{U}(x).$$

Proof 1. This follows from 1.8.1 and 1.8.36.

2. By complete distributivity we have

$$\begin{aligned}\inf_{\mathcal{U}\in\mathbf{U}(\mathcal{F})} \lambda\mathcal{U}(x) &= \inf_{\mathcal{U}\in\mathbf{U}(\mathcal{F})} \sup_{U\in\mathcal{U}} \delta(x,U)\\ &= \sup_{\sigma\in\prod_{\mathcal{U}\in\mathbf{U}(\mathcal{F})}\mathcal{U}} \inf_{\mathcal{U}\in\mathbf{U}(\mathcal{F})} \delta(x,\sigma(\mathcal{U})).\end{aligned}$$

From 1.2.2 it follows that, for any $\sigma \in \prod_{\mathcal{U}\in\mathbf{U}(\mathcal{F})} \mathcal{U}$, we can find a finite subcollection $\mathbf{U}_\sigma \subset \mathbf{U}(\mathcal{F})$ such that

$$\bigcup_{\mathcal{U}\in\mathbf{U}_\sigma} \sigma(\mathcal{U}) \in \mathcal{F}.$$

Making use of this, it follows that

$$\begin{aligned}\inf_{\mathcal{U}\in\mathbf{U}(\mathcal{F})} \lambda\mathcal{U}(x) &\leq \sup_{\sigma\in\prod_{\mathcal{U}\in\mathbf{U}(\mathcal{F})}\mathcal{U}} \inf_{\mathcal{U}\in\mathbf{U}_\sigma} \delta(x,\sigma(\mathcal{U}))\\ &= \sup_{\sigma\in\prod_{\mathcal{U}\in\mathbf{U}(\mathcal{F})}\mathcal{U}} \delta\left(x, \bigcup_{\mathcal{U}\in\mathbf{U}_\sigma} \sigma(\mathcal{U})\right)\\ &\leq \alpha\mathcal{F}(x).\end{aligned}$$

The other inequality is immediate.

□

By fine-tuning the proof of the second property in 1.8.37 we can show that the infimum is a minimum.

1.8.38 Proposition *Let (X,δ) be an approach space. For all $\mathcal{F}\in\mathbf{F}(X)$ and $x \in X$ there exists $\mathcal{U} \in \mathbf{U}(\mathcal{F})$ such that $\alpha\mathcal{F}(x) = \lambda\mathcal{U}(x)$.*

Proof Let $x \in X$ and suppose on the contrary that, for all $\mathcal{U} \in \mathbf{U}(\mathcal{F})$, $\alpha\mathcal{F}(x) < \lambda\mathcal{U}(x)$. Then it follows from 1.8.36 and the definition of the adherence operator that, for each $\mathcal{U} \in \mathbf{U}(\mathcal{F})$, there exists $U_\mathcal{U} \in \mathcal{U}$ such that $\alpha\mathcal{F}(x) < \delta(x,U_\mathcal{U})$. It now follows from 1.2.2 that there exists a finite subset $\mathbf{U}_0 \subset \mathbf{U}(\mathcal{F})$ such that $\bigcup_{\mathcal{U}\in\mathbf{U}_0} U_\mathcal{U} \in \mathcal{F}$ and consequently, by definition of the adherence operator,

$$\begin{aligned}\alpha\mathcal{F}(x) &\geq \delta\left(x, \bigcup_{\mathcal{U}\in\mathbf{U}_0} U_\mathcal{U}\right)\\ &= \min_{\mathcal{U}\in\mathbf{U}_0} \delta(x,U_\mathcal{U})\\ &> \alpha\mathcal{F}(x),\end{aligned}$$

which is a contradiction.

□

The foregoing results generalize the facts that in a topological space a filter converges to a point if and only if all ultrafilters which are finer converge to that point and that a filter adheres to a point if and only if there exists an ultrafilter which is finer and which converges to that point.

The following characterizations of the adherence operator in terms of approach systems and gauges will also be of use.

1.8.39 Proposition *Let (X, δ) be an approach space. Then, for all $\mathcal{F} \in \mathbf{F}(X)$ and $x \in X$, we have*

$$\begin{aligned} \alpha\mathcal{F}(x) &= \sup_{\varphi \in \mathcal{A}(x)} \sup_{F \in \mathcal{F}} \inf_{y \in F} \varphi(y) \\ &= \sup_{d \in \mathcal{G}} \sup_{F \in \mathcal{F}} \inf_{y \in F} d(x, y). \end{aligned}$$

Proof This follows from 1.8.24, 1.8.5, and the definition of adherence operator given in 1.8.34.

□

1.8.40 Proposition *Let (X, δ) be an approach space. Then, for all $\mathcal{F} \in \mathbf{F}(X)$ and $x, y \in X$, we have*

$$\lambda\mathcal{F}(x) \leq \delta(x, \{y\}) + \lambda\mathcal{F}(y).$$

Proof Let $x, y \in X$; then, by 1.8.31, we have

$$\begin{aligned} \lambda\mathcal{F}(x) &= \sup_{d \in \mathcal{G}} \inf_{F \in \mathcal{F}} \sup_{z \in F} d(x, z) \\ &\leq \sup_{d \in \mathcal{G}} \inf_{F \in \mathcal{F}} \sup_{z \in F} (d(x, y) + d(y, z)) \\ &\leq \sup_{d \in \mathcal{G}} d(x, y) + \sup_{d \in \mathcal{G}} \inf_{F \in \mathcal{F}} \sup_{z \in F} d(y, z) \\ &= \delta(x, \{y\}) + \lambda\mathcal{F}(y). \end{aligned}$$

□

For an example showing that the inequality in the foregoing proposition is both best possible and strict in general we refer to 4.2.2.

To conclude this section, for easy reference, we recall the transition formulas which we proved. The tables also contain some formulas which we did not prove and we leave the easy verifications to the reader. Furthermore, sometimes we also give an alternative way of writing some of these transitions. For a justification of these alternatives we refer to 3.1.10. It follows from 1.8.26 and 1.8.27 that the formulas proved in 1.8.5, 1.8.13, 1.8.15, 1.8.24, 1.8.28, 1.8.30, 1.8.31, and 1.8.39 remain valid when either an approach system is replaced by a basis or a gauge is replaced by a basis. The transition from distances to hull operators given below is not the one which we used in 1.8.15, since there we went via the gauge associated with a distance, but rather it is the notationally somewhat complicated but conceptually simple formula which we essentially obtained earlier in the section on hull operators.

Transition formulas from a distance δ

Limit operator (1.8.1)	$\lambda\mathcal{F}(x) = \sup_{A\in\sec(\mathcal{F})} \delta(x, A)$
Adherence operator (1.8.34)	$\alpha\mathcal{F}(x) = \sup_{F\in\mathcal{F}} \delta(x, F)$
Approach system (1.8.23)	$\mathcal{A}(x) = \left\{\varphi \in [0,\infty]^X \mid \forall A \subset X : \inf_{y\in A} \varphi(y) \leq \delta(x, A)\right\}$
Gauge (1.8.3)	$\mathcal{G} = \left\{d \in pq\mathbf{M}^\infty(X) \mid \forall A \subset X : \inf_{a\in A} d(\cdot, a) \leq \delta_A\right\}$ or $\mathcal{G} = \{d \in pq\mathbf{M}^\infty(X) \mid \delta_d \leq \delta\}$ (see 3.1.10)
Tower (1.8.19)	$\mathfrak{t}_\epsilon(A) = A^{(\epsilon)} = \{x \in X \mid \delta(x, A) \leq \epsilon\}$
Hull operator (1.6.6 and 1.8.16)	$h(\mu)(x) = \sup_{\omega<\infty} \sup_{\epsilon>0} \inf_{i=1}^{n(\omega,\epsilon)} (m_i^{\omega,\epsilon} + \delta(x, M_i^{\omega,\epsilon}))$ where, for each $\omega < \infty$, $\left(\inf_{i=1}^{n(\omega,\epsilon)} (m_i^{\omega,\epsilon} + \theta_{M_i^{\omega,\epsilon}})\right)_{\epsilon>0}$ is a development for $\mu \wedge \omega$

Transition formulas from a limit operator λ

Distance (1.8.2)	$\delta(x, A) = \inf_{\mathcal{U} \in \mathbf{U}(A)} \lambda\mathcal{U}(x)$
Adherence operator (1.8.37)	$\alpha\mathcal{F}(x) = \inf_{\mathcal{U} \in \mathbf{U}(\mathcal{F})} \lambda\mathcal{U}(x)$
Approach system	$\mathcal{A}(x) = \left\{\varphi \in [0, \infty]^X \mid \forall \mathcal{U} \in \mathbf{U}(X) : \sup_{U \in \mathcal{U}} \inf_{y \in U} \varphi(y) \leq \lambda\mathcal{U}(x)\right\}$
Gauge	$\mathcal{G} = \left\{d \in pq\mathbf{M}^\infty(X) \mid \forall \mathcal{U} \in \mathbf{U}(X) : \sup_{U \in \mathcal{U}} \inf_{y \in U} d(\cdot, y) \leq \lambda\mathcal{U}\right\}$ or $\mathcal{G} = \{d \in pq\mathbf{M}^\infty(X) \mid \lambda_d \leq \lambda\}$ (see 3.1.10)
Tower	$\mathfrak{t}_\epsilon(A) = \{x \in X \mid \exists \mathcal{F} \in \mathbf{F}(A) : \lambda\mathcal{F}(x) \leq \epsilon\}$

Transition formulas from an approach system $\mathcal{A}$

Distance (1.8.24)	$\delta(x, A) = \sup_{\varphi \in \mathcal{A}(x)} \inf_{y \in A} \varphi(y)$
Gauge (1.8.11)	$\mathcal{G} = \{d \in pq\mathbf{M}^\infty(X) \mid \forall x \in X : d(x, \cdot) \in \mathcal{A}(x)\}$
Limit operator (1.8.30)	$\lambda\mathcal{F}(x) = \sup_{\varphi \in \mathcal{A}(x)} \inf_{F \in \mathcal{F}} \sup_{y \in F} \varphi(y)$
Adherence operator (1.8.39)	$\alpha\mathcal{F}(x) = \sup_{\varphi \in \mathcal{A}(x)} \sup_{F \in \mathcal{F}} \inf_{y \in F} \varphi(y)$
Hull operator (1.8.28)	$h(\mu)(x) = \sup_{\varphi \in \mathcal{A}(x)} \inf_{y \in X} (\mu + \varphi)(y)$

Transition formulas from a gauge $\mathcal{G}$

Distance (1.8.5)	$\delta(x,A) = \sup_{d\in\mathcal{G}} \inf_{y\in A} d(x,y)$ or $\delta = \sup_{d\in\mathcal{G}} \delta_d$ (see 3.1.10)
Approach system (1.8.10)	$\mathcal{A}(x) = \left\{\varphi \in [0,\infty]^X \mid \{d(x,\cdot) \mid d \in \mathcal{G}\} \text{ dominates } \varphi\right\}$
Limit operator (1.8.31)	$\lambda\mathcal{F}(x) = \sup_{d\in\mathcal{G}} \inf_{F\in\mathcal{F}} \sup_{y\in F} d(x,y)$ or $\lambda = \sup_{d\in\mathcal{G}} \lambda_d$ (see 3.1.10)
Adherence operator (1.8.39)	$\alpha\mathcal{F}(x) = \sup_{d\in\mathcal{G}} \sup_{F\in\mathcal{F}} \inf_{y\in F} d(x,y)$ or $\alpha = \sup_{d\in\mathcal{G}} \alpha_d$ (see 3.1.10)
Hull operator (1.8.15)	$h(\mu)(x) = \sup_{d\in\mathcal{G}} \inf_{y\in X} (\mu(y) + d(x,y))$ or $h = \sup_{d\in\mathcal{G}} h_d$ (see 3.1.10)

Transition formulas from a regular function frame $\mathcal{R}$

Distance	$\delta(x,A) = \sup\left\{\rho(x) \mid \rho \in \mathcal{R}, \rho_{\vert A} = 0\right\}$
Hull operator (1.8.21)	$h(\mu) = \sup\{\nu \in \mathcal{R} \mid \nu \leq \mu\}$

Transition formulas from a tower t

Distance (1.8.20)	$\delta(x, A) = \inf\{\epsilon \in \mathbb{R}^+ \mid x \in t_\epsilon(A)\}$
Limit operator	$\lambda\mathcal{F}(x) = \sup_{A \in \sec(\mathcal{F})} \inf\{\epsilon \in \mathbb{R}^+ \mid x \in t_\epsilon(A)\}$
Adherence operator	$\alpha\mathcal{F}(x) = \sup_{F \in \mathcal{F}} \inf\{\epsilon \in \mathbb{R}^+ \mid x \in t_\epsilon(F)\}$
Approach system	$\mathcal{A}(x) = \left\{\varphi \in [0,\infty]^X \mid \forall A \subset X, \forall \epsilon > 0 : x \in t_\epsilon(A) \Rightarrow \inf_{y \in A} \varphi(y) \leq \epsilon\right\}$
Gauge	$\mathcal{G} = \left\{d \in pq\mathbf{M}^\infty(X) \mid \forall A \subset X :\ t_\epsilon(A) \subset \left\{\inf_{y \in A} d(\cdot, y) \leq \epsilon\right\}\right\}$ or $\mathcal{G} = \left\{d \in pq\mathbf{M}^\infty(X) \mid\ t_\epsilon \leq t_\epsilon^d\right\}$ (see 3.1.10)

Transition formulas from a hull operator h

Distance (1.8.17)	$\delta(x, A) = h(\theta_A)(x)$
Approach system (1.8.25)	$\mathcal{A}(x) = \left\{\varphi \in [0,\infty]^X \mid \forall \mu \in [0,\infty]^X : \inf_{y \in X} (\mu + \varphi)(y) \leq h(\mu)(x)\right\}$
Gauge	$\mathcal{G} = \left\{d \in pq\mathbf{M}^\infty(X) \mid \forall \mu \in [0,\infty]^X : \inf_{y \in X} (\mu(y) + d(\cdot, y)) \leq h(\mu)\right\}$ or $\mathcal{G} = \{d \in pq\mathbf{M}^\infty(X) \mid\ h_d \leq h\}$ (see 3.1.10)
Regular function frame (1.8.22)	$\mathcal{R} = \left\{\mu \in [0,\infty]^X \mid h(\mu) = \mu\right\}$

1.9 CONTRACTIONS

In this section we introduce the morphisms which are most naturally associated with the structures defined in the foregoing sections. We will give the definition in terms of distances, but we will immediately show that these morphisms can be nicely characterized by all of the other approach structures as well.

1.9.1 Definition If (X, δ) and (X', δ') are approach spaces, then a function

$$f : X \longrightarrow X'$$

is called a *contraction* if

$$\forall x \in X,\ \forall A \in 2^X : \ \delta'(f(x), f(A)) \leq \delta(x, A).$$

In order to formulate the equivalent condition involving the hull operator we need to introduce a supplementary notation. If $f : X \to X'$ is a function and $\mu \in [0, \infty]^X$, then we define and denote the function

$$f(\mu) : X' \longrightarrow [0, \infty]$$

by

$$f(\mu)(x') \doteqdot \inf_{x \in f^{-1}(x')} \mu(x) \qquad \forall x' \in X'.$$

1.9.2 Theorem *If (X, δ) and (X', δ') are approach spaces and $f : X \longrightarrow X'$, then the following are equivalent:*

1. *f is a contraction,*
2. $\forall \mathcal{F} \in \mathbf{F}(X) : \ \lambda'(\text{stack} f(\mathcal{F})) \circ f \leq \lambda \mathcal{F}$,
3. $\forall \mathcal{F} \in \mathbf{U}(X) : \ \lambda'(\text{stack} f(\mathcal{F})) \circ f \leq \lambda \mathcal{F}$,
4. $\forall x \in X, \forall \varphi' \in \mathcal{A}'(f(x)) : \ \varphi' \circ f \in \mathcal{A}(x)$,
5. $\forall d' \in \mathcal{G}' : \ d' \circ (f \times f) \in \mathcal{G}$,
6. $\forall A \in 2^X, \forall \epsilon \in [0, \infty] : \ f(\mathfrak{t}_\epsilon(A)) \subset \mathfrak{t}'_\epsilon(f(A))$,
7. $\forall \mu \in [0, \infty]^X : \ h'(f(\mu)) \leq f(h(\mu))$,
8. $\forall \nu \in \mathcal{R}' : \nu \circ f \in \mathcal{R}$.

Proof 1 ⇒ 2. For any $\mathcal{F} \in \mathbf{F}(X)$, we have

$$\begin{aligned}\lambda'(\operatorname{stack} f(\mathcal{F})) \circ f &= \sup_{W \in \sec(\operatorname{stack} f(\mathcal{F}))} \delta'_W \circ f \\ &\leq \sup_{U \in \sec(\mathcal{F})} \delta'_{f(U)} \circ f \\ &\leq \sup_{U \in \sec(\mathcal{F})} \delta_U \\ &= \lambda \mathcal{F}.\end{aligned}$$

2 ⇒ 3. This is trivial.

3 ⇒ 1. For any $A \in 2^X$, we have

$$\begin{aligned}\delta'_{f(A)} \circ f &= \inf_{\mathcal{W} \in \mathbf{U}(f(A))} \lambda' \mathcal{W} \circ f \\ &\leq \inf_{\mathcal{U} \in \mathbf{U}(A)} \lambda'(\operatorname{stack} f(\mathcal{U})) \circ f \\ &\leq \inf_{\mathcal{U} \in \mathbf{U}(A)} \lambda \mathcal{U} \\ &= \delta_A.\end{aligned}$$

6 ⇒ 1. For any $x \in X$ and $A \subset X$, we have $x \in A^{(\delta(x,A))}$, and thus $f(x) \in f(A)^{(\delta(x,A))'}$ which precisely means that $\delta'(f(x), f(A)) \leq \delta(x, A)$.

1 ⇒ 4. Suppose on the contrary that there exist $x_0 \in X$, $\varphi'_0 \in \mathcal{A}'(f(x_0))$, $\epsilon_0 > 0$, and $\omega_0 < \infty$ such that, for all $\varphi \in \mathcal{A}(x_0)$,

$$A(\varphi) \doteq \{x \in X \mid \varphi'_0(f(x)) \wedge \omega_0 > \varphi(x) + \epsilon\}$$

is nonempty. Then we obtain

$$\begin{aligned}\sup_{\varphi \in \mathcal{A}(x_0)} \delta(x_0, A(\varphi)) \wedge \omega_0 &= \sup_{\varphi \in \mathcal{A}(x_0)} \sup_{\psi \in \mathcal{A}(x_0)} \inf_{x \in A(\varphi)} \psi(x) \wedge \omega_0 \\ &\leq \sup_{\varphi \in \mathcal{A}(x_0)} \sup_{\psi \in \mathcal{A}(x_0)} \inf_{x \in A(\varphi \vee \psi)} \varphi \vee \psi(x) \wedge \omega_0 \\ &= \sup_{\varphi \in \mathcal{A}(x_0)} \inf_{x \in A(\varphi)} \varphi(x) \wedge \omega_0 \\ &\leq \sup_{\varphi \in \mathcal{A}(x_0)} \inf_{x \in A(\varphi)} \varphi'_0(f(x)) \wedge \omega_0 - \epsilon_0 \\ &\leq \sup_{\varphi \in \mathcal{A}(x_0)} \sup_{\xi \in \mathcal{A}'(f(x_0))} \inf_{y \in f(A(\varphi))} \xi(y) \wedge \omega_0 - \epsilon_0 \\ &= \sup_{\varphi \in \mathcal{A}(x_0)} \delta'(f(x_0), f(A(\varphi))) \wedge \omega_0 - \epsilon_0,\end{aligned}$$

which is impossible.

$4 \Rightarrow 5$. Let $d' \in \mathcal{G}'$. Then it follows that, for all $x \in X$, $d'(f(x),\cdot) \in \mathcal{A}'(f(x))$ and hence $d'(f(x),\cdot) \circ f \in \mathcal{A}(x)$. Consequently $d' \circ (f \times f) \in \mathcal{G}$.

$5 \Rightarrow 7$. Let $\mu \in [0,\infty]^X$ and $x' \in X'$. If $x' \notin f(X)$, then there is nothing to prove. Suppose therefore that $x' \in f(X)$ and fix $x \in f^{-1}(x')$. Then we obtain

$$\begin{aligned} h'(f(\mu))(x') &= \sup_{d'\in\mathcal{G}'} \inf_{u'\in X'} \left(\inf_{u\in f^{-1}(u')} \mu(u) + d'(x',u') \right) \\ &\le \sup_{d'\in\mathcal{G}'} \inf_{v\in X} \left(\inf_{u\in f^{-1}(f(v))} \mu(u) + d'(x',f(v)) \right) \\ &\le \sup_{d\in\mathcal{G}} \inf_{v\in X} (\mu(v) + d(x,v)) \\ &= h(\mu)(x), \end{aligned}$$

which from the arbitrariness of $x \in f^{-1}(x')$ proves our claim.

$7 \Rightarrow 6$. Let $x \in \mathrm{t}_\epsilon(A)$; then $h(\theta_A)(x) = \delta_A(x) \le \epsilon$. Consequently it follows that

$$\begin{aligned} \delta'_{f(A)}(f(x)) &= h'\big(\theta_{f(A)}\big)(f(x)) \\ &\le f(h(\theta_A))(f(x)) \\ &= \inf_{z\in f^{-1}(f(x))} h(\theta_A)(z) \\ &\le h(\theta_A)(x) \\ &\le \epsilon, \end{aligned}$$

which proves the assertion.

$7 \Rightarrow 8$. Let $\nu \in \mathcal{R}'$; then

$$\begin{aligned} \nu &= h'(\nu) \\ &\le h'(f(\nu\circ f)) \\ &\le f(h(\nu\circ f)), \end{aligned}$$

and it follows that $\nu \circ f \le h(\nu \circ f)$, i.e. $\nu \circ f \in \mathcal{R}$.

$8 \Rightarrow 7$. Let $\mu \in [0,\infty]^X$; then

$$\big(h'(f(\mu))\big) \circ f \le f(\mu) \circ f \le \mu$$

and since $(h'(f(\mu))) \circ f \in \mathcal{R}$, it follows that also $(h'(f(\mu))) \circ f \le h(\mu)$ and thus $h'(f(\mu)) \le f(h(\mu))$.

□

It should be pointed out that if in the fourth characterization of the foregoing proposition, the approach system $(\mathcal{A}'(x'))_{x' \in X'}$ is replaced by a basis or if in the fifth characterization, similarly, the gauge $\mathcal{G}'$ is replaced by a basis, then the results remain valid.

Some interesting examples of contractions are given in the following propositions, where the space $\mathbb{P}$ introduced in 1.8.33 plays the main role.

1.9.3 Proposition *Let (X, δ) be an approach space. Then, for any $A \in 2^X$, the distance functional*

$$\delta_A : (X, \delta) \longrightarrow \mathbb{P} : x \longrightarrow \delta(x, A)$$

is a contraction.

Proof Let $x \in X$ and $B \subset X$. If $B = \emptyset$, then

$$\delta_{\mathbb{P}}(\delta_A(x), \delta_A(B)) = \delta(x, B) = \infty.$$

Otherwise, since

$$\delta_{\mathbb{P}}(\delta_A(x), \delta_A(B)) = \left(\delta(x, A) - \sup_{b \in B} \delta(b, A) \right) \vee 0,$$

the result is an immediate consequence of 1.1.2.

□

1.9.4 Proposition *Let (X, δ) be an approach space with regular function frame $\mathcal{R}$. Then, for any function $\xi \in [0, \infty]^X$, the following are equivalent:*

1. *$\xi \in \mathcal{R}$, i.e. ξ is regular,*
2. *$\xi : (X, \delta) \longrightarrow \mathbb{P}$ is a contraction.*

Proof Both the first and the second property can be expressed as local properties of ξ. By 1.8.22 and 1.8.28, $\xi \in \mathcal{R}$ if and only if

$$\forall x \in \{\xi < \infty\}, \forall \epsilon > 0\, \exists \varphi \in \mathcal{A}(x)\, \forall y \in X : \; \xi(x) \leq \xi(y) + \varphi(y) + \epsilon$$

and (*)

$$\forall x \in \{\xi = \infty\}, \forall \omega < \infty\, \exists \varphi \in \mathcal{A}(x)\, \forall y \in X : \; \omega \leq \xi(y) + \varphi(y).$$

From 1.9.2 and the description of the approach system of $\mathbb{P}$ it follows that $\xi : (X,\delta) \to \mathbb{P}$ is a contraction if and only if

$$\forall x \in \{\xi < \infty\}, \forall \epsilon > 0, \forall \omega < \infty\ \exists \varphi \in \mathcal{A}(x)\ \forall y \in \{\xi < \xi(x)\} :$$

$$(\xi(x) - \xi(y)) \wedge \omega \leq \varphi(y) + \epsilon$$

and (**)

$$\forall x \in \{\xi = \infty\}, \forall \epsilon > 0, \forall \omega < \infty, \forall \sigma < \infty\ \exists \varphi \in \mathcal{A}(x) :$$

$$(\theta_{]\sigma,\infty]} \circ \xi) \wedge \omega \leq \varphi + \epsilon.$$

First, let us consider the case $x \in \{\xi < \infty\}$. Then in the first condition of (**) it clearly suffices to consider $\omega = \xi(x)$ and thus this condition is equivalent with the first condition of (*). Next, let us consider the case where $\xi(x) = \infty$. Clearly, in the second condition of (**) it suffices to consider $\omega = \sigma$ and then it is easily seen that this condition is equivalent with

$$\forall x \in \{\xi = \infty\}, \forall \epsilon > 0, \forall \omega < \infty\ \exists \varphi \in \mathcal{A}(x)\ \forall y \in \{\xi \leq \omega\} :\ \omega \leq \varphi(y) + \epsilon,$$

which, taking ω and 2ω for ϵ and ω, is further seen to be equivalent with

$$\forall x \in \{\xi = \infty\}, \forall \omega < \infty\ \exists \varphi \in \mathcal{A}(x)\ \forall y \in \{\xi \leq \omega\} :\ \omega \leq \varphi(y).$$

This condition clearly implies the second condition of (*). Conversely the second condition of (*) implies that

$$\forall x \in \{\xi = \infty\}, \forall \omega < \infty\ \exists \varphi \in \mathcal{A}(x)\ \forall y \in X :\ 2\omega \leq \xi(y) + \varphi(y),$$

which clearly implies the above equivalent form of the second condition of (**).

□

1.9.5 Corollary *Let (X,δ) be an approach space. Then, for any $A \in 2^X$, we have that $\delta_A \in \mathcal{R}$, i.e. δ_A is a regular function.*

Proof This follows at once from 1.9.3 and 1.9.4.

□

1.9.6 Corollary *Let (X,δ) be an approach space. Then, for any $\mathcal{F} \in \mathbf{F}(X)$, we have that $\lambda\mathcal{F} \in \mathcal{R}$ and $\alpha\mathcal{F} \in \mathcal{R}$, i.e. $\lambda\mathcal{F}$ and $\alpha\mathcal{F}$ are regular functions.*

Proof This follows at once from the fact that $\mathcal{R}$ is closed under the formation of arbitrary suprema, 1.8.1, 1.8.34, and 1.9.5.

□

1.10 THE TOPOLOGICAL CONSTRUCT **AP**

That approach spaces form the objects and contractions form the morphisms of a construct is easily verified. We will denote this construct by **AP**. For the fundamental theory of topological constructs we refer the reader to Appendix A or to Adámek *et al.* [1].

1.10.1 Theorem **AP** *is a topological construct. In particular, given approach spaces* $(X_j, \delta_j)_{j\in J}$, *consider the source*

$$(f_j : X \longrightarrow (X_j, \delta_j))_{j\in J}$$

in **AP** *and suppose that, for each* $j \in J$, $(\mathcal{B}_j(x))_{x\in X_j}$ *is a basis for the approach system in* X_j. *Then a basis for the initial approach system on* X *is given by*

$$\mathcal{B}(x) \doteqdot \left\{ \sup_{j\in K} \xi_j \circ f_j \mid K \in 2^{(J)},\ \forall j \in K : \xi_j \in \mathcal{B}_j(f_j(x)) \right\} \qquad \forall x \in X.$$

Proof Clearly **AP** is concrete, small-fibered, and fulfills the terminal separator property. The main property we have to show is the existence of initial structures in **AP**. Let X be a set and let J be a class. For each $j \in J$, let (X_j, δ_j) be an approach space, let $(\mathcal{B}_j(x))_{x\in X_j}$ be a basis for the approach system $(\mathcal{A}_j(x))_{x\in X_j}$ in X_j, and let $f_j : X \longrightarrow X_j$ be a function. For each $x \in X$, let $\mathcal{B}(x)$ be defined as above. It is easily seen that $\mathcal{B}(x)$ is a basis for an ideal and that it fulfills (B1). To prove that $(\mathcal{B}(x))_{x\in X}$ fulfills (B2), let $\sup_{j\in K} \xi_j \circ f_j \in \mathcal{B}(x)$ and let $\epsilon > 0$ and $\omega < \infty$. For each $j \in K$, there is a family $\left(\xi^j_z\right)_{z\in X_j} \in \prod_{z\in X_j} \mathcal{B}_j(z)$ such that, for all $z, y \in X_j$,

$$\xi^j_{f_j(x)}(z) + \xi^j_z(y) + \epsilon \geq \xi_j(y) \wedge \omega.$$

For each $t \in X$, let

$$\nu_t \doteqdot \sup_{j\in K} \xi^j_{f_j(t)} \circ f_j.$$

Then $\nu_t \in \mathcal{B}(t)$ and, for any $t, s \in X$, we have

$$\begin{aligned}\left(\sup_{j\in K} \xi_j \circ f_j\right)(s) \wedge \omega &= \sup_{j\in K} \xi_j(f_j(s)) \wedge \omega \\ &\leq \sup_{j\in K} \left(\xi^j_{f_j(x)}(f_j(t)) + \xi^j_{f_j(t)}(f_j(s)) + \epsilon\right) \\ &\leq \nu_x(t) + \nu_t(s) + \epsilon.\end{aligned}$$

This proves that $(\mathcal{A}(x))_{x\in X} \doteqdot \left(\widehat{\mathcal{B}(x)}\right)_{x\in X}$ is indeed an approach system on X. By construction, all the functions f_j, $j \in J$, are contractions. To prove that $(\mathcal{A}(x))_{x\in X}$ is an initial structure, let $\left(Z, (\mathcal{C}(z))_{z\in Z}\right)$ be an approach space determined by its approach system and let

$$g : Z \longrightarrow X$$

be such that all compositions

$$f_j \circ g : \left(Z, (\mathcal{C}(z))_{z\in Z}\right) \longrightarrow \left(X_j, (\mathcal{A}_j(x))_{x\in X_j}\right)$$

are contractions. Further, let $z \in Z$ and let $\sup_{j\in K} \xi_j \circ f_j \in \mathcal{A}(g(z))$. Since

$$\left(\sup_{j\in K} \xi_j \circ f_j\right) \circ g = \sup_{j\in K} \xi_j \circ (f_j \circ g)$$

and since, for all $j \in K$,

$$\xi_j \circ (f_j \circ g) \in \mathcal{C}(z),$$

it follows from 1.9.2 that g is a contraction.

□

The following results are immediate consequences of the fact that **AP** is a topological construct (Appendix A and Adámek *et al.* [1]).

1.10.2 Corollary *1. Both initial and final structures exist in* **AP**.

2. For any set X, the **AP***-fiber, i.e. the set of all approach structures on X, is a complete lattice.*

3. Constant functions are contractions in **AP**.

□

1.10.3 Proposition *1. Given a set X, the finest (discrete) approach structure on it is determined by any (and all) of the following structures:*

1. *Distance:* $\delta : X \times 2^X \longrightarrow [0,\infty]$ *where, for all* $x \in X$ *and* $A \subset X$,
$$\delta(x,A) = \begin{cases} 0 & x \in A, \\ \infty & x \notin A. \end{cases}$$
2. *Limit:* $\lambda : \mathbf{F}(X) \longrightarrow [0,\infty]^X$ *where, for all* $x \in X$ *and* $\mathcal{F} \in \mathbf{F}(X)$,
$$\lambda\mathcal{F} = \begin{cases} \theta_{\{x\}} & \mathcal{F} = \mathrm{stack}x, \\ \infty & \mathcal{F} \neq \mathrm{stack}x. \end{cases}$$
3. *Approach system:* $(\mathcal{A}(x))_{x\in X}$ *where, for all* $x \in X$,
$$\mathcal{A}(x) = \left\{\varphi \in [0,\infty]^X \mid \varphi(x) = 0\right\}.$$
4. *Gauge:* $\mathcal{G} = pq\mathbf{M}^\infty(X)$.
5. *Tower:* $(\mathfrak{t}_\epsilon)_{\epsilon\in\mathbb{R}^+}$ *where, for all* $\epsilon \in \mathbb{R}^+$ *and* $A \subset X$,
$$\mathfrak{t}_\epsilon(A) = A.$$
6. *Hull operator:* $h : [0,\infty]^X \longrightarrow [0,\infty]^X$ *where, for all* $\mu \in [0,\infty]^X$,
$$h(\mu) = \mu.$$
7. *Regular function frame:* $\mathcal{R} = [0,\infty]^X$.

2. Given a set X, the coarsest (indiscrete or trivial) approach structure on it is determined by any (and all) of the following structures:

1. *Distance:* $\delta : X \times 2^X \longrightarrow [0,\infty]$ *where, for all* $x \in X$ *and* $A \subset X$,
$$\delta(x,A) = \begin{cases} 0 & A \neq \emptyset, \\ \infty & A = \emptyset. \end{cases}$$
2. *Limit:* $\lambda : \mathbf{F}(X) \longrightarrow [0,\infty]^X$ *where, for all* $\mathcal{F} \in \mathbf{F}(X)$,
$$\lambda\mathcal{F} = 0.$$
3. *Approach system:* $(\mathcal{A}(x))_{x\in X}$ *where, for all* $x \in X$,
$$\mathcal{A}(x) = \{0\}.$$
4. *Gauge:* $\mathcal{G} = \{0\}$.
5. *Tower:* $(\mathfrak{t}_\epsilon)_{\epsilon\in\mathbb{R}^+}$ *where, for all* $\epsilon \in \mathbb{R}^+$ *and* $A \subset X$,
$$\mathfrak{t}_\epsilon(A) = \begin{cases} X & A \neq \emptyset, \\ \emptyset & A = \emptyset. \end{cases}$$
6. *Hull operator:* $h : [0,\infty]^X \longrightarrow [0,\infty]^X$ *where, for all* $\mu \in [0,\infty]^X$,
$$h(\mu) = \inf \mu.$$
7. *Regular function frame:* $\mathcal{R} = \left\{\varphi \in [0,\infty]^X \mid \varphi \textit{ constant}\right\}$.

Proof This is a straightforward verification which we leave to the reader.

□

It is not possible to give characterizations of initial and final structures in **AP** for all the different defining concepts of approach spaces. Actually the situation is analogous to the one in **TOP**. However, there are a few important constructions which we have to explain, namely initial structures by means of gauges and limit operators. Furthermore, for completeness and since they are so simple, we will also give the descriptions of both initial and final structures by means of regular function frames.

1.10.4 Theorem *Given approach spaces $(X_j, \mathcal{G}_j)_{j\in J}$, defined by means of their gauges, consider the source*

$$(f_j : X \longrightarrow (X_j, \mathcal{G}_j))_{j\in J}$$

in **AP**. *If, for each $j \in J$, $\mathcal{H}_j$ is a basis for the gauge $\mathcal{G}_j$, then a basis for the initial gauge on X is given by*

$$\mathcal{H} \doteqdot \left\{ \sup_{j\in K} d_j \circ (f_j \times f_j) \mid K \in 2^{(J)},\ \forall j \in K : d_j \in \mathcal{H}_j \right\}.$$

Proof This goes along the same lines as 1.10.1.

□

1.10.5 Theorem *Given approach spaces $(X_j, \lambda_j)_{j\in J}$, defined by means of their limit operators, consider the source*

$$(f_j : X \longrightarrow (X_j, \lambda_j))_{j\in J}$$

in **AP**. *Then the initial limit operator on X is given by*

$$\lambda\mathcal{F} = \sup_{j\in J} \lambda_j(\text{stack} f_j(\mathcal{F})) \circ f_j \qquad \forall \mathcal{F} \in \mathbf{F}(X).$$

Proof It suffices to restrict our attention to ultrafilters since the general case then immediately follows upon applying 1.8.37. We denote by λ the initial limit operator on X, by $\mathcal{A}$ the approach system associated with λ, and by $\mathcal{A}_j$ the approach system associated with λ_j, for each $j \in J$.

Now let $\mathcal{U}$ be an ultrafilter on X and let $x \in X$ be arbitrary. We know from 1.10.1 that

$$\begin{aligned}\lambda\mathcal{U}(x) &= \sup_{\varphi\in\mathcal{A}(x)} \sup_{U\in\mathcal{U}} \inf_{y\in U} \varphi(y) \\ &= \sup_{J_0\in 2^{(J)}} \sup_{\varphi\in\prod\limits_{j\in J_0} \mathcal{A}_j(f_j(x))} \sup_{U\in\mathcal{U}} \inf_{y\in U} \sup_{j\in J_0} \varphi_j \circ f_j(y).\end{aligned}$$

Furthermore let $J_0 \in 2^{(J)}$, $\varphi \in \prod\limits_{j\in J_0} \mathcal{A}_j(f_j(x))$, and $U \in \mathcal{U}$ be fixed. Then it follows that

$$\begin{aligned}\inf_{y\in U} \sup_{j\in J_0} \varphi_j \circ f_j(y) &= \sup_{t\in J_0^U} \inf_{y\in U} \varphi_{t(y)} \circ f_{t(y)}(y) \\ &= \sup_{t\in J_0^U} \inf_{j\in J_0} \inf_{y\in t^{-1}(j)} \varphi_j \circ f_j(y).\end{aligned} \tag{*}$$

Since $\mathcal{U}$ is an ultrafilter, it follows that, for each $t \in J_0^U$, there exists $j_t \in J_0$ such that $t^{-1}(j_t) \in \mathcal{U}$ and consequently we have

$$\begin{aligned}\sup_{t\in J_0^U} \inf_{j\in J_0} \inf_{y\in t^{-1}(j)} \varphi_j \circ f_j(y) &\leq \sup_{t\in J_0^U} \inf_{y\in t^{-1}(j_t)} \varphi_{j_t} \circ f_{j_t}(y) \\ &\leq \sup_{t\in J_0^U} \sup_{W\in\mathcal{U}} \inf_{y\in W} \varphi_{j_t} \circ f_{j_t}(y) \\ &= \sup_{W\in\mathcal{U}} \sup_{j\in J_0} \inf_{y\in W} \varphi_j \circ f_j(y).\end{aligned} \tag{**}$$

Because this inequality holds for all choices of $U \in \mathcal{U}$ it follows from (*) and (**) that

$$\sup_{U\in\mathcal{U}} \inf_{y\in U} \sup_{j\in J_0} \varphi_j \circ f_j(y) \leq \sup_{W\in\mathcal{U}} \sup_{j\in J_0} \inf_{y\in W} \varphi_j \circ f_j(y).$$

Since, as is immediately clear, the other inequality also holds we can now incorporate this in our previous calculation for $\lambda\mathcal{U}(x)$ in order to obtain

$$\begin{aligned}\lambda\mathcal{U}(x) &= \sup_{J_0\in 2^{(J)}} \sup_{\varphi\in\prod_{j\in J_0} \mathcal{A}_j(f_j(x))} \sup_{U\in\mathcal{U}} \inf_{y\in U} \sup_{j\in J_0} \varphi_j\circ f_j(y)\\ &= \sup_{J_0\in 2^{(J)}} \sup_{\varphi\in\prod_{j\in J_0} \mathcal{A}_j(f_j(x))} \sup_{U\in\mathcal{U}} \sup_{j\in J_0} \inf_{y\in U} \varphi_j\circ f_j(y)\\ &= \sup_{J_0\in 2^{(J)}} \sup_{\varphi\in\prod_{j\in J_0} \mathcal{A}_j(f_j(x))} \sup_{j\in J_0} \sup_{U\in\mathcal{U}} \inf_{y\in f_j(U)} \varphi_j(y)\\ &= \sup_{j\in J} \sup_{\varphi\in\mathcal{A}_j(f_j(x))} \sup_{U\in\mathcal{U}} \inf_{y\in f_j(U)} \varphi_j(y)\\ &= \sup_{j\in J} \lambda_j(\text{stack} f_j\mathcal{U})(f_j(x)),\end{aligned}$$

which proves our claim.

□

1.10.6 Theorem *Given approach spaces $(X_j, \mathcal{R}_j)_{j\in J}$, defined by means of their regular function frames, consider the sink*

$$(f_j : (X_j, \mathcal{R}_j) \longrightarrow X)_{j\in J}$$

in **AP**. *Then the final regular function frame on X is given by*

$$\mathcal{R} \doteqdot \left\{\mu \in [0,\infty]^X \mid \forall j \in J : \ \mu\circ f_j \in \mathcal{R}_j\right\}.$$

Proof It is straightforward to verify that $\mathcal{R}$ as defined above is indeed a regular function frame and by construction it is of course the finest one turning all the maps f_j, $j \in J$, into contractions.

□

In the following theorem, given a set of functions $\mathcal{B} \subset [0,\infty]^X$, we denote by $\mathcal{B}^\wedge$ the set of all infima of finite sets of members of $\mathcal{B}$ and by $\mathcal{B}^\vee$ the set of all suprema of arbitrary families of members of $\mathcal{B}$.

1.10.7 Theorem *Given approach spaces* $(X_j, \mathcal{R}_j)_{j\in J}$, *defined by means of their regular function frames, consider the source*

$$(f_j : X \longrightarrow (X_j, \mathcal{R}_j))_{j\in J}$$

in **AP**. *Then the initial regular function frame on* X *is given by*

$$\mathcal{R} \doteqdot \{\mu \circ f_j \mid j \in J,\ \mu \in \mathcal{R}_j\}^{\wedge\bigvee}.$$

Proof Since translation commutes with infima, suprema, and composition, it is at once clear by construction that $\mathcal{R}$ as defined above is indeed a regular function frame and that it is the coarsest one turning all the maps $f_j,\ j \in J$, into contractions.

□

We end this section by showing that **AP**, as **TOP**, is simply generated, i.e. it has an *initially dense object.*

1.10.8 Proposition $\mathbb{P}$ *is initially dense in* **AP**. *More precisely, for any approach space* (X, δ), *the source*

$$(\delta_A : (X, \delta) \longrightarrow \mathbb{P})_{A\in 2^X}$$

is initial.

Proof From 1.9.3 we already know that if δ_{in} stands for the initial distance, then $\delta_{in} \le \delta$. Conversely, if λ (respectively $\lambda_{in}, \lambda_{\mathbb{P}}$) stands for the limit operator associated with δ (respectively $\delta_{in}, \delta_{\mathbb{P}}$), then it follows from 1.10.5 that, for any $\mathcal{U} \in \mathbf{U}(X)$,

$$\begin{aligned}
\lambda_{in}\mathcal{U}(x) &= \sup_{A\in 2^X} \lambda_{\mathbb{P}}(\delta_A(\mathcal{U}))(\delta_A(x)) \\
&= \sup_{A\in 2^X} \sup_{U\in\mathcal{U}} \delta_{\mathbb{P}}(\delta_A(x), \delta_A(U)) \\
&\ge \sup_{U\in\mathcal{U}} \delta_{\mathbb{P}}(\delta_U(x), \delta_U(U)) \\
&\ge \sup_{U\in\mathcal{U}} \delta_U(x) \\
&= \lambda\mathcal{U}(x),
\end{aligned}$$

which proves our claim.

□

1.10.9 Corollary *For any approach space* (X, δ), *the source*

$$(\xi : (X, \delta) \longrightarrow \mathbb{P})_{\xi \in \mathcal{R}}$$

is initial.

Proof This follows at once from 1.9.4 and 1.10.8. □

2
TOPOLOGICAL APPROACH SPACES

In this chapter we will see that topological spaces can be viewed as special types of approach spaces and that, more precisely, **TOP** can actually be embedded as a full and isomorphism-closed subconstruct of **AP**. As far as notation is concerned, from now on, whenever we say that $(X, \mathcal{T})$ is a topological space, $\mathcal{T}$ will stand for the collection of open sets. Structures derived from $\mathcal{T}$, such as the associated closure operator, will be denoted, for example, by $\mathrm{cl}_{\mathcal{T}}$. If no confusion can arise we may also drop reference to $\mathcal{T}$ altogether.

2.1 TOPOLOGICAL APPROACH STRUCTURES

Given a topological space $(X, \mathcal{T})$, we associate with it a natural approach space in the following way. We define the function

$$\delta_{\mathcal{T}} : X \times 2^X \longrightarrow [0, \infty]$$

by

$$\delta_{\mathcal{T}}(x, A) \doteqdot \begin{cases} 0 & x \in \mathrm{cl}_{\mathcal{T}}(A), \\ \infty & x \notin \mathrm{cl}_{\mathcal{T}}(A). \end{cases}$$

2.1.1 Proposition *Suppose $(X, \mathcal{T}) \in |\mathbf{TOP}|$. Then the function*

$$\delta_{\mathcal{T}} : X \times 2^X \longrightarrow [0, \infty]$$

is a distance on X.

Proof (D1) and (D2) are immediate. (D3) follows from the fact that, for any $A, B \subset X$, we have $\mathrm{cl}_{\mathcal{T}}(A \cup B) = \mathrm{cl}_{\mathcal{T}}(A) \cup \mathrm{cl}_{\mathcal{T}}(B)$. (D4) follows from the fact that, for all $\epsilon < \infty$, $A^{(\epsilon)} = \mathrm{cl}_{\mathcal{T}}(A)$ and $A^{(\infty)} = X$.

□

2.1.2 Definition An approach space of type $(X, \delta_{\mathcal{T}})$, for some topology $\mathcal{T}$ on X, will be called a *topological approach space*. A distance of type $\delta_{\mathcal{T}}$ will be called a *topological distance*.

This way of looking at a topological space involves of course no more than a change of language. Instead of saying that a point is in the closure of a set, we say it has zero distance to the set. Saying that the point is not in the closure of a set amounts to saying it has an infinite distance to that set.

In what follows, if $\mathcal{F}$ is a filter in a topological space X, then we denote by $\mathrm{adh}\mathcal{F}$ the set of adherence points of $\mathcal{F}$ and by $\lim \mathcal{F}$ the set of limit points of $\mathcal{F}$. If $x \in X$, then we denote the fact that $\mathcal{F}$ adheres to x by $\mathcal{F} \dashv x$ and the fact that $\mathcal{F}$ converges to x by $\mathcal{F} \to x$. If necessary, in order to avoid confusion, we may add subscripts or superscripts to these notations.

2.1.3 Proposition *Suppose* $(X, \mathcal{T}) \in |\mathbf{TOP}|$. *Then the adherence and limit operator in* $(X, \delta_{\mathcal{T}})$ *are determined by:*

1. $\alpha_{\mathcal{T}}\mathcal{F} = \theta_{\mathrm{adh}\mathcal{F}} \qquad \forall \mathcal{F} \in \mathbf{F}(X),$
2. $\lambda_{\mathcal{T}}\mathcal{F} = \theta_{\lim \mathcal{F}} \qquad \forall \mathcal{F} \in \mathbf{F}(X).$

Proof We give the proof for the adherence operator; the one for the limit operator is precisely the same. It follows from 1.8.34 that, for any $\mathcal{F} \in \mathbf{F}(X)$,

$$\begin{aligned} \alpha_{\mathcal{T}}\mathcal{F} &= \sup_{F \in \mathcal{F}} (\delta_{\mathcal{T}})_F \\ &= \sup_{F \in \mathcal{F}} \theta_{\mathrm{cl}_{\mathcal{T}}(F)} \\ &= \theta_{\bigcap_{F \in \mathcal{F}} \mathrm{cl}_{\mathcal{T}}(F)} \\ &= \theta_{\mathrm{adh}\mathcal{F}}. \end{aligned}$$

□

Again, the interpretation of the foregoing result is clear. The distance a point is away from being a limit point of a filter is either zero or infinite, depending on whether it is a topological limit point or not. Notice that this again confirms the plausibility of our intuitive interpretation of limit and adherence in approach spaces. A limit operator of type $\lambda_{\mathcal{T}}$ will be called a *topological limit operator*.

In the sequel usc will stand for upper semicontinuous and lsc for lower semicontinuous. We recall that a function $\mu : (X, \mathcal{T}) \to ([0,\infty], \mathcal{T}_{\mathbb{E}})$ is lower semicontinuous if and only if $\mu : (X, \mathcal{T}) \to ([0,\infty], \mathcal{T}_{\mathbb{P}})$ is continuous. Here $\mathcal{T}_{\mathbb{E}}$ stands for the usual topology on $[0,\infty]$ and $\mathcal{T}_{\mathbb{P}} \doteqdot \{]a,\infty] \mid a \in [0,\infty]\} \cup \{[0,\infty]\}$ (see also 2.2.9).

2.1.4 Proposition *Suppose* $(X, \mathcal{T}) \in |\mathbf{TOP}|$. *Then the approach system in* $(X, \delta_{\mathcal{T}})$ *is given by*

$$\mathcal{A}_{\mathcal{T}}(x) \doteqdot \left\{\varphi \in [0,\infty]^X \mid \varphi(x) = 0,\ \varphi \text{ usc in } x\right\} \qquad \forall x \in X$$

and a basis for this system is given by

$$\mathcal{B}_{\mathcal{T}}(x) \doteqdot \{\theta_V \mid V \in \mathcal{V}_{\mathcal{T}}(x)\} \qquad \forall x \in X,$$

where $(\mathcal{V}_{\mathcal{T}}(x))_{x\in X}$ *is the neighborhood system in* $(X, \mathcal{T})$.

Proof From 1.8.23 it follows that

$$\mathcal{A}_{\mathcal{T}}(x) \doteqdot \left\{\varphi \in [0,\infty]^X \mid \forall A \subset X : x \in \mathrm{cl}_{\mathcal{T}}(A) \Rightarrow \inf_{a\in A} \varphi(a) = 0\right\} \qquad \forall x \in X.$$

If $\varphi \in \mathcal{A}_{\mathcal{T}}(x)$, then, since $x \in \mathrm{cl}_{\mathcal{T}}(\{x\})$, it follows that $\varphi(x) = 0$. Now let $\alpha > 0$. If $x \in \mathrm{cl}_{\mathcal{T}}(\{\varphi \geq \alpha\})$, then

$$\inf_{a\in\{\varphi\geq\alpha\}} \varphi(a) = 0,$$

which is absurd. Hence, for all $\alpha > 0$, $x \in \mathrm{int}_{\mathcal{T}}(\{\varphi < \alpha\})$, which proves that φ is upper semicontinuous in x. The converse follows from the fact that if φ is upper semicontinuous in x and $\varphi(x) = 0$, then, for any $\alpha > 0$, $\{\varphi < \alpha\}$ is a neighborhood of x.

That $\mathcal{B}_{\mathcal{T}}(x)$ is a basis for $\mathcal{A}_{\mathcal{T}}(x)$ follows easily from the definitions.

□

In the foregoing result we discovered a new concept characterizing topological spaces, namely local distances, which turn out to be well-known types of functions. Of course this is only new in form, but not in essence. As the theorem shows, the essential information is contained in a canonical basis for the local distances, which essentially turns out to be the neighborhood system. An approach system of type $\mathcal{A}_{\mathcal{T}}$ will be called a *topological approach system*.

2.1.5 Proposition *Suppose $(X, \mathcal{T}) \in |\mathbf{TOP}|$. Then the gauge in $(X, \delta_{\mathcal{T}})$ is given by*

$$\mathcal{G}_{\mathcal{T}} \doteqdot \{d \in pq\mathbf{M}^{\infty}(X) \mid \mathcal{T}_d \subset \mathcal{T}\},$$

where $\mathcal{T}_d$ stands for the topology generated by d.

Proof From 1.8.3 we obtain that

$$\mathcal{G} = \{d \in pq\mathbf{M}^{\infty}(X) \mid x \in \mathrm{cl}_{\mathcal{T}}(A) \Rightarrow \delta_d(x, A) = 0\}$$

from which the result immediately follows.

□

2.1.6 Proposition *Suppose $(X, \mathcal{T}) \in |\mathbf{TOP}|$. Then the tower in $(X, \delta_{\mathcal{T}})$ is given by the family $\left(\mathrm{t}_\epsilon^{\mathcal{T}}\right)_{\epsilon \in \mathbb{R}^+}$ where*

$$\mathrm{t}_\epsilon^{\mathcal{T}} : 2^X \longrightarrow 2^X : A \longrightarrow \mathrm{cl}_{\mathcal{T}}(A) \qquad \forall \epsilon \in \mathbb{R}^+.$$

Proof For any $\epsilon \in \mathbb{R}^+$ and $A \subset X$, we have

$$\begin{aligned} \mathrm{t}_\epsilon^{\mathcal{T}}(A) &= \{x \in X \mid \delta_{\mathcal{T}}(x, A) \leq \epsilon\} \\ &= \{x \in X \mid \delta_{\mathcal{T}}(x, A) = 0\} \\ &= \mathrm{cl}_{\mathcal{T}}(A). \end{aligned}$$

□

The pretopological structures of a tower in **TOP** are all equal to the given topological structure. So here too, as was to be expected, no new structure is involved. A tower of type $\mathrm{t}^{\mathcal{T}}$ will be called a *topological tower.*

2.1.7 Proposition *Suppose $(X, \mathcal{T}) \in |\mathbf{TOP}|$. Then the hull operator in $(X, \delta_{\mathcal{T}})$ is given by*

$$h_{\mathcal{T}}(\mu)(x) \doteqdot \sup_{V \in \mathcal{V}_{\mathcal{T}}(x)} \inf_{y \in V} \mu(y) \qquad \forall x \in X,\ \forall \mu \in [0, \infty]^X,$$

where $\mathcal{V}_{\mathcal{T}}(x)$ stands for the neighborhood filter in x. Alternatively, $h_{\mathcal{T}}(\mu)$ is also characterized as being the greatest lower semicontinuous function smaller than μ, called its lower semicontinuous regularization.

Proof The formula follows from 1.8.28. The alternative description is well-known and can be found for instance in Bourbaki [17].

□

A hull operator of type $h_{\mathcal{T}}$ will be called a *topological hull operator*.

2.1.8 Proposition *Suppose $(X, \mathcal{T}) \in |\mathbf{TOP}|$. Then the regular function frame in $(X, \delta_{\mathcal{T}})$ is given by*

$$\mathcal{R}_{\mathcal{T}} \doteqdot \left\{\mu \in [0,\infty]^X \mid \mu \text{ lsc}\right\}.$$

Proof This follows from 1.8.22 and 2.1.7.

□

Again, both the lower semicontinuous regularization and the collection of lower semicontinuous functions on a topological space are well-known concepts used in various areas of mathematics. The novelty which we find here is that they can both serve as an alternative primitive concept to define a topology. A regular function frame of type $\mathcal{R}_{\mathcal{T}}$ will be called a *topological regular function frame*.

The foregoing results show that we can identify topological spaces with certain types of approach spaces. The next result gives an internal characterization of these spaces.

2.1.9 Theorem *An approach space (X, δ) is topological if and only if*

$$\delta\left(X \times 2^X\right) = \{0, \infty\}.$$

Proof The only-if part follows from the definition of a topological approach space. To show the if part it suffices to note that if, for all $A \subset X$, we put $\mathrm{cl}(A) \doteqdot \{x \in X \mid \delta(x, A) = 0\}$, then cl is a topological closure operator and δ is the associated distance.

□

The result of 2.1.9, in combination with 1.10.3, shows that, on any set, both the finest and the coarsest approach structures are topological.

2.2 THE EMBEDDING OF TOP IN AP

In the foregoing section we have seen that a topological space can be viewed as a special type of approach space. That, moreover, **TOP** is concretely embedded in **AP** is a consequence of the following proposition.

2.2.1 Proposition *Let $(X, \mathcal{T})$ and $(X', \mathcal{T}')$ be topological spaces and consider the function $f : X \longrightarrow X'$. Then the following are equivalent:*

1. $f : (X, \mathcal{T}) \longrightarrow (X', \mathcal{T}')$ *is continuous,*
2. $f : (X, \delta_\mathcal{T}) \longrightarrow (X', \delta_{\mathcal{T}'})$ *is a contraction.*

Proof If $A \subset X$, then it follows from 2.1.6 that

$$f(\mathrm{cl}_\mathcal{T} A) \subset \mathrm{cl}_{\mathcal{T}'}(f(A)) \Leftrightarrow \forall \epsilon \in \mathbb{R}^+ : f\left(A^{(\epsilon)}\right) \subset (f(A))^{(\epsilon)'},$$

which by 1.9.2 proves our claim.

□

2.2.2 Corollary *The functor*

$$\begin{array}{ccc} \mathbf{TOP} & \rightarrow & \mathbf{AP} \\ (X, \mathcal{T}) & \mapsto & (X, \delta_\mathcal{T}) \\ f & \mapsto & f \end{array}$$

is a full concrete embedding of **TOP** *into* **AP**.

□

We will now prove that **TOP** is actually very nicely embedded in **AP**. In contrast to most known topological constructs which do not have subconstructs which are simultaneously bireflectively and bicoreflectively embedded, such as for instance **TOP** itself, we will prove that **TOP** is simultaneously bireflectively and bicoreflectively embedded in **AP**. For categorical background on these notions we refer to Appendix A or to Adámek *et al.* [1].

We recall that two important aspects of the fact that **TOP** is bireflectively embedded in **AP** are: (1) for each approach structure, there exists a finest coarser topological structure on the same underlying set and (2) initial structures of topological spaces are the same whether they are taken in **TOP** or in **AP**.

2.2.3 Proposition *For any space* $(X, \delta) \in |\mathbf{AP}|$, *the function* $\mathrm{cl} : 2^X \longrightarrow 2^X$, *defined by*

$$\mathrm{cl}(A) \doteqdot \{x \in X \mid \delta(x, A) < \infty\} \qquad \forall A \in 2^X,$$

is a pretopological closure operator.

Proof This follows from (D1), (D2), and (D3).

□

We recall that the category **PRTOP** of pretopological spaces and continuous maps is a topological category in which **TOP** is bireflectively embedded. For more information on **PRTOP** we refer the reader to Appendix B or to Choquet [25] and Lowen-Colebunders [84], for example.

2.2.4 Theorem **TOP** *is embedded as a concretely bireflective subconstruct of* **AP**. *For any space* $(X, \delta) \in |\mathbf{AP}|$, *its* **TOP**-*bireflection is given by*

$$id_X : (X, \delta) \longrightarrow (X, \delta^{tr}),$$

where δ^{tr} *is the distance associated with the topological bireflection of the pretopological closure operator* cl.

Proof The topological bireflection of this closure operator is obtained by a standard transfinite process which produces a topological closure operator, cl_{tr}. To verify that $id_X : (X, \delta) \to (X, \delta^{tr})$ is a contraction it suffices to note that if $\delta(x, A) < \infty$, for some $x \in X$ and $A \subset X$, then $\delta^{tr}(x, A) = 0$. Now suppose that $(Y, \mathcal{T}) \in |\mathbf{TOP}|$ and that

$$f : (X, \delta) \longrightarrow (Y, \delta_{\mathcal{T}})$$

is a contraction. Then, for any $x \in X$ and $A \subset X$, we have

$$\begin{aligned} x \in \mathrm{cl}(A) &\Rightarrow \delta(x, A) < \infty \\ &\Rightarrow \delta_{\mathcal{T}}(f(x), f(A)) < \infty \\ &\Rightarrow \delta_{\mathcal{T}}(f(x), f(A)) = 0 \\ &\Rightarrow f(x) \in \mathrm{cl}_{\mathcal{T}}(f(A)). \end{aligned}$$

Hence $f : (X, \mathrm{cl}) \longrightarrow (Y, \mathrm{cl}_{\mathcal{T}})$ is continuous as a function between pretopological spaces. It then follows that also $f : (X, \mathrm{cl}_{tr}) \longrightarrow (Y, \mathrm{cl}_{\mathcal{T}})$ is continuous as a function between topological spaces, which in turn, by 2.2.1, means that $f : (X, \delta^{tr}) \longrightarrow (Y, \delta_{\mathcal{T}})$ is a contraction.

□

2.2.5 Corollary **TOP** *is closed under the formation of limits and initial structures in* **AP**. *In particular, a product in* **AP** *of a family of topological approach spaces is a topological approach space and, likewise, a subspace in* **AP** *of a topological approach space is a topological approach space.*

Proof This follows from 2.2.4 and Adámek *et al.* [1].

□

Although, as the previous results show, it is important to know from a structural point of view that **TOP** is bireflectively embedded in **AP**, we will not often have recourse to considering the **TOP**-bireflection of an approach space. It is easily seen that if the distance is finite, then the **TOP**-bireflection is indiscrete. This shows that it will usually not be a very interesting topology to consider. The situation becomes totally different for the dual property, bicoreflectivity, as we will now see. Here too we refer to Appendix A and to Adámek *et al.* [1] for detailed information. We also recall that two important aspects of the fact that **TOP** is bicoreflectively embedded in **AP** are: (1) for each approach structure there exists a coarsest finer topological structure on the same underlying set and (2) final structures of topological spaces are the same whether they are taken in **TOP** or in **AP**.

2.2.6 Theorem **TOP** *is embedded as a concretely bicoreflective subconstruct of* **AP**. *For any space* $(X,\delta) \in |\mathbf{AP}|$, *its* **TOP**-*bicoreflection is given by*

$$id_X : \left(X, \delta^{tc}\right) \longrightarrow (X,\delta),$$

where δ^{tc} *is the distance associated with the topological closure operator given by*

$$\mathrm{cl}_\delta(A) \doteqdot \{x \in X \mid \delta(x,A) = 0\} \qquad \forall A \in 2^X.$$

Proof It is easily verified that cl_δ is indeed a topological closure operator and that $id_X : \left(X, \delta^{tc}\right) \to (X,\delta)$ is a contraction. Now suppose that $(Y,\mathcal{T}) \in |\mathbf{TOP}|$ and that

$$f : (Y, \delta_{\mathcal{T}}) \longrightarrow (X,\delta)$$

is a contraction. Then, for any $x \in Y$ and $A \subset Y$ such that $x \in \mathrm{cl}_{\mathcal{T}}(A)$, we have

$$\delta(f(x), f(A)) \leq \delta_{\mathcal{T}}(x,A) = 0,$$

which proves that

$$f : (Y, \delta_{\mathcal{T}}) \longrightarrow \left(X, \delta^{tc}\right)$$

is also a contraction.

□

2.2.7 Corollary $\mathbf{TOP}$ *is closed under the formation of colimits and final structures in* $\mathbf{AP}$*. In particular, a coproduct in* $\mathbf{AP}$ *of a family of topological approach spaces is a topological approach space and, likewise, a quotient in* $\mathbf{AP}$ *of a topological approach space is a topological approach space.*

Proof This follows from 2.2.6 and Adámek *et al.* [1].

□

In the next chapter we will discuss the importance of the $\mathbf{TOP}$-bicoreflection of an approach space. Anticipating this, we will now describe this bicoreflection by means of the most important other basic structures. Given an approach space (X, δ), we will denote the topology underlying the topological bicoreflection by $\mathcal{T}_\delta$.

2.2.8 Proposition *For any space* $(X, \delta) \in |\mathbf{AP}|$*, the following hold:*

1. *Convergence in* $\mathcal{T}_\delta$ *is characterized by*

$$\mathcal{F} \to x \Leftrightarrow \lambda\mathcal{F}(x) = 0 \qquad \forall x \in X,\ \forall \mathcal{F} \in \mathbf{F}(X).$$

2. *The neighborhoods in* $\mathcal{T}_\delta$ *are characterized by*

$$\mathcal{V}(x) = \left\{V \in 2^X \mid \exists \epsilon > 0, \exists \varphi \in \mathcal{B}(x) :\ \{\varphi < \epsilon\} \subset V\right\} \qquad \forall x \in X$$

and by

$$\mathcal{V}(x) = \left\{V \in 2^X \mid \exists \epsilon > 0, \exists d \in \mathcal{H} :\ B_d(x, \epsilon) \subset V\right\} \qquad \forall x \in X,$$

where $(\mathcal{B}(x))_{x \in X}$ *is a basis for the approach system and where* $\mathcal{H}$ *is a basis for the gauge.*

Proof The proof of the first property follows from

$$\begin{aligned} \mathcal{F} \to x &\Leftrightarrow x \in \bigcap_{A \in \mathrm{sec}(\mathcal{F})} \mathrm{cl}_\delta(A) \\ &\Leftrightarrow \forall A \in \mathrm{sec}(\mathcal{F}) :\ \delta(x, A) = 0 \\ &\Leftrightarrow \sup_{A \in \mathrm{sec}(\mathcal{F})} \delta(x, A) = 0 \\ &\Leftrightarrow \lambda\mathcal{F}(x) = 0. \end{aligned}$$

The proof of the second property for the approach system follows from

$$\begin{aligned}
V \in \mathcal{V}(x) &\Leftrightarrow x \notin \mathrm{cl}_\delta(X \setminus V) \\
&\Leftrightarrow \exists \epsilon > 0 : \sup_{\varphi \in \mathcal{B}(x)} \inf_{y \in X \setminus V} \varphi(y) > \epsilon \\
&\Leftrightarrow \exists \epsilon > 0,\ \exists \varphi \in \mathcal{B}(x) : \inf_{y \in X \setminus V} \varphi(y) > \epsilon \\
&\Leftrightarrow \exists \epsilon > 0,\ \exists \varphi \in \mathcal{B}(x) : \{\varphi < \epsilon\} \subset V
\end{aligned}$$

and analogously for the gauge.

□

2.2.9 Examples We refer to the two examples which we considered in 1.8.33.

The distance of the first example is $\delta_{\mathbb{E}} : [0, \infty] \times 2^{[0,\infty]} \longrightarrow [0, \infty]$, where

$$\delta_{\mathbb{E}}(x, A) \doteqdot \begin{cases} 0 & x = \infty, A \text{ unbounded}, \\ \infty & x = \infty, A \text{ bounded}, \\ \inf_{a \in A \cap \mathbb{R}^+} |x - a| & x < \infty. \end{cases}$$

The topological bicoreflection of this space is determined by $([0, \infty], \mathcal{T}_{\mathbb{E}})$, where $\mathcal{T}_{\mathbb{E}}$ is the topology of the *Alexandroff compactification* of $[0, \infty[$, i.e. the usual (Euclidean) topology.

Notice that this distance has a remarkable property. Although the Alexandroff compactification of $[0, \infty[$ is metrizable, it is of course not metrizable by a metric which extends the usual Euclidean metric. The distance $\delta_{\mathbb{E}}$, however, is the unique distance which "distancizes" the topology of the Alexandroff compactification and which extends the usual metric.

The distance of the second example is $\delta_{\mathbb{P}} : [0, \infty] \times 2^{[0,\infty]} \longrightarrow [0, \infty]$, where

$$\delta_{\mathbb{P}}(x, A) \doteqdot \begin{cases} (x - \sup A) \vee 0 & A \neq \emptyset, \\ \infty & A = \emptyset. \end{cases}$$

The topological bicoreflection of this space is determined by $([0, \infty], \mathcal{T}_{\mathbb{P}})$, where

$$\mathcal{T}_{\mathbb{P}} \doteqdot \{]a, \infty] \mid a \in [0, \infty]\} \cup \{[0, \infty]\}.$$

2.2.10 Proposition *Let (X,δ) be an approach space. Then, for any $\xi \in \mathcal{R}$,*

$$\xi : (X, \mathcal{T}_\delta) \longrightarrow ([0,\infty], \mathcal{T}_{\mathbb{P}})$$

is a continuous map, in particular:

1. *for any $A \in 2^X$, the distance functional*

$$\delta_A : (X, \mathcal{T}_\delta) \longrightarrow ([0,\infty], \mathcal{T}_{\mathbb{P}})$$

is a continuous map,

2. *for any $\mathcal{F} \in \mathbf{F}(X)$, both*

$$\lambda\mathcal{F} : (X, \mathcal{T}_\delta) \longrightarrow ([0,\infty], \mathcal{T}_{\mathbb{P}})$$

and

$$\alpha\mathcal{F} : (X, \mathcal{T}_\delta) \longrightarrow ([0,\infty], \mathcal{T}_{\mathbb{P}})$$

are continuous maps.

Proof This follows at once from 1.9.4, 1.9.6, 2.2.6, and 2.2.9.

□

If, in the foregoing result, we replace the topology $\mathcal{T}_{\mathbb{P}}$ by the Euclidean topology $\mathcal{T}_{\mathbb{E}}$, then all maps are lower semicontinuous.

2.3 BIREFLECTIVE BICOREFLECTIVE SUBCONSTRUCTS

It is well known that in many familiar topological constructs, such as **TOP**, **PRTOP**, and **UNIF**, there are no nontrivial (simultaneously) bireflective (and) bicoreflective subconstructs, Kannan [59], Hušek [54]. From the foregoing section we know that for approach spaces the situation is different, since **TOP** is both bireflectively and bicoreflectively embedded in **AP**. Actually, **TOP** is not the only such subconstruct. In the present section we will completely describe all subconstructs of **AP** which are simultaneously bireflectively and bicoreflectively embedded in it.

Before stating the results, we need to recall some items from Herrlich and Lowen [50], where a technique for determining bireflective bicoreflective subconstructs of specific topological constructs was given. The fundamental result of that paper determines a relationship between bireflective bicoreflective subconstructs on the one hand and particular function algebras on the other hand, and it was pointed out that **AP** is a construct of the right type.

Whenever necessary and without introducing another notation for it, we consider $[0,\infty]$ to be an *algebra* (in the sense of the aforementioned paper) equipped with addition, with bounded subtraction, with the operation of taking finite infima, and with the operation of taking arbitrary suprema. More precisely, we consider $[0,\infty]$ to be equipped with the following class of operations:

1. supremum for arbitrary sets S: $\vee_S : \mathcal{A}^S \to \mathcal{A} \;:\; (a_s)_{s\in S} \to \sup_{s\in S} a_s$,
2. infimum for finite sets S: $\wedge_S : \mathcal{A}^S \to \mathcal{A} \;:\; (a_s)_{s\in S} \to \min_{s\in S} a_s$,
3. addition for $a \in [0,\infty]$: $\oplus_a : \mathcal{A} \to \mathcal{A} \;:\; b \to b + a$,
4. bounded subtraction for $a \in [0,\infty]$: $\ominus_a : \mathcal{A} \to \mathcal{A} \;:\; b \to (b-a) \vee 0$.

Since this basic algebra is thus completely fixed, and contrary to the notations used in Herrlich and Lowen [50], we will usually drop any reference to it in our notations.

For any set X, the set $[0,\infty]^X$, equipped with the pointwise extensions of the above operations, then also becomes an algebra.

By a *subalgebra* of a given algebra we mean a subset which is closed under the above class of operations. It follows from 1.7.1 that subalgebras of $[0,\infty]^X$ and regular function frames on X are the same notions and it is precisely this fact that ensures that the techniques of [50] can be applied to **AP**. More precisely it was shown that **AP** is isomorphic to $\mathbf{T}([0,\infty])$ for the above algebra $[0,\infty]$. $\mathbf{T}([0,\infty])$ is the topological construct with objects pairs $(X,\mathcal{A})$, where $\mathcal{A}$ is a subalgebra of $[0,\infty]^X$ and with morphisms those functions between the underlying sets which preserve the subalgebras under composition. We refer the reader to [50] for more details.

If $\Gamma \subset [0,\infty]^{[0,\infty]}$, then we will denote by $\widetilde{\Gamma}$ the smallest *monoidal subalgebra* of $[0,\infty]^{[0,\infty]}$ containing Γ, i.e. the smallest subset of $[0,\infty]^{[0,\infty]}$ which contains Γ and which is simultaneously a *submonoid* of $\left([0,\infty]^{[0,\infty]}, \circ, id_{[0,\infty]}\right)$ and a *subalgebra* of $[0,\infty]^{[0,\infty]}$.

If $\mathcal{C}$ is a subconstruct of **AP**, then we define

$$\Omega(\mathcal{C}) \doteqdot \left\{\gamma \in [0,\infty]^{[0,\infty]} \mid \forall (X,\mathcal{R}) \in |\mathcal{C}|, \forall \mu \in \mathcal{R} : \gamma \circ \mu \in \mathcal{R}\right\}.$$

If $\Gamma \subset [0,\infty]^{[0,\infty]}$, then $\mathcal{C}$ is called Γ-*invariant* if

$$\forall (X,\mathcal{R}) \in |\mathcal{C}|, \forall \gamma \in \Gamma, \forall \mu \in \mathcal{R} : \gamma \circ \mu \in \mathcal{R},$$

i.e. if $\Gamma \subset \Omega(\mathcal{C})$.

Further, we also recall the definition of the following full subconstructs of **AP** (or equivalently of $\mathbf{T}([0,\infty])$), determined by an arbitrary subset $\Gamma \subset [0,\infty]^{[0,\infty]}$:

$$\mathbf{T}([0,\infty],\Gamma) \doteqdot \{(X,\mathcal{R}) \in |\mathbf{AP}| \mid (\gamma \in \Gamma \text{ and } \mu \in \mathcal{R}) \Rightarrow \gamma \circ \mu \in \mathcal{R}\}.$$

As we have already mentioned, we will always write $\mathbf{T}(\Gamma)$ instead of $\mathbf{T}([0,\infty],\Gamma)$. We then call $\mathcal{C}$ *maximal* Γ-*invariant* if $\mathcal{C} = \mathbf{T}(\Gamma)$.

Note that $\Gamma \subset \Omega(\mathcal{C})$ is equivalent with $|\mathcal{C}| \subset |\mathbf{T}(\Gamma)|$.

Finally, we recall that

$$\Gamma_{\mathbf{AP}} \doteqdot \left\{\mu \in [0,\infty]^{[0,\infty]} \mid \mu \text{ nonexpansive, increasing, and } \mu(\infty) = \sup_{x<\infty} \mu(x)\right\}$$

and

$$\begin{aligned}\Gamma_{\mathbf{TOP}} &\doteqdot \left\{\mu \in [0,\infty]^{[0,\infty]} \mid \mu \text{ lsc and increasing}\right\}\\ &= \left\{\mu \in [0,\infty]^{[0,\infty]} \mid \mu \text{ left continuous and increasing}\right\}\end{aligned}$$

are monoidal subalgebras of $[0,\infty]^{[0,\infty]}$ and thus also regular function frames on $[0,\infty]$ and that we have

$$\mathbf{AP} = \mathbf{T}(\Gamma_{\mathbf{AP}}) \text{ and } \mathbf{TOP} = \mathbf{T}(\Gamma_{\mathbf{TOP}}).$$

We will now tag all the bireflective bicoreflective subconstructs of **AP**. More precisely we will prove that there exists a bijection between the class of these subconstructs on the one hand and a particular set of monoidal subalgebras of $[0,\infty]^{[0,\infty]}$ on the other hand.

In the following proposition we gather together some basic results concerning the concepts which we have introduced. The easy proof of this proposition is left to the reader.

2.3.1 Proposition *1. If $\mathcal{C}$ is a subconstruct of* **AP**, *then $\Omega(\mathcal{C})$ is the final regular function frame on $[0,\infty]$ for the sink*

$$(\mu : (X,\mathcal{R}) \longrightarrow [0,\infty])_{(X,\mathcal{R})\in|\mathcal{C}|,\mu\in\mathcal{R}}$$

and, moreover, it is a monoidal subalgebra of $[0,\infty]^{[0,\infty]}$.

2. If $(\mathcal{C}_j)_{j\in J}$ is a family of subconstructs of **AP**, *then $\Omega\left(\bigcup_{j\in J} \mathcal{C}_j\right) = \bigcap_{j\in J} \Omega(\mathcal{C}_j)$.*

3. If $\Gamma \subset \Lambda \subset [0,\infty]^{[0,\infty]}$, then $\mathbf{T}(\Lambda)$ is a subconstruct of $\mathbf{T}(\Gamma)$.

4. If $(\Gamma_j)_{j\in J}$ is a family of subsets of $[0,\infty]^{[0,\infty]}$, then we have that

$$\left|\mathbf{T}\left(\bigcup_{j\in J}\Gamma_j\right)\right| = \bigcap_{j\in J} |\mathbf{T}(\Gamma_j)|.$$

Proof This follows by straightforward verification.

□

2.3.2 Proposition *For any $\Gamma \subset [0,\infty]^{[0,\infty]}$, we have:*

1. $\mathbf{T}(\Gamma) = \mathbf{T}\left(\widetilde{\Gamma}\right)$,

2. $\widetilde{\Gamma} = \Omega(\mathbf{T}(\Gamma))$.

Proof 1. Since $\Gamma \subset \widetilde{\Gamma}$, we have $\left|\mathbf{T}\left(\widetilde{\Gamma}\right)\right| \subset |\mathbf{T}(\Gamma)|$. On the other hand, we have that $\Gamma \subset \Omega(\mathbf{T}(\Gamma))$ and it follows from 2.3.1 that the right hand side is a monoidal subalgebra of $[0,\infty]^{[0,\infty]}$. Hence, it also follows that $\widetilde{\Gamma} \subset \Omega(\mathbf{T}(\Gamma))$ and that $|\mathbf{T}(\Gamma)| \subset \left|\mathbf{T}\left(\widetilde{\Gamma}\right)\right|$ and we are done.

2. It follows from the first part that $\widetilde{\Gamma} \subset \Omega(\mathbf{T}(\Gamma))$. Conversely, since $\left([0,\infty],\widetilde{\Gamma}\right) \in |\mathbf{T}(\Gamma)|$, it follows from 2.3.1 that we also have $\Omega(\mathbf{T}(\Gamma)) \subset \widetilde{\Gamma}$.

□

2.3.3 Proposition *For any subconstruct $\mathcal{C}$ of* **AP** *the following properties hold:*

1. $\mathcal{C}$ is a subconstruct of $\mathbf{T}(\Omega(\mathcal{C}))$,

2. $\mathcal{C}$ is maximal Γ-invariant, for some Γ, if and only if $\mathcal{C} = \mathbf{T}(\Omega(\mathcal{C}))$ and in that case $\Gamma = \Omega(\mathcal{C})$,

3. $\Omega(\mathbf{T}(\Omega(\mathcal{C}))) = \Omega(\mathcal{C})$.

Proof The first part follows immediately from the definitions. To prove the second part, let $\mathcal{C} \doteqdot \mathbf{T}(\Gamma)$. Since $\Gamma \subset \Omega(\mathcal{C})$, it follows from 2.3.1 that $\mathbf{T}(\Omega(\mathcal{C}))$ is a subconstruct of $\mathcal{C}$. The third part follows from 2.3.2 and the fact that $\Omega(\mathcal{C})$ is a monoidal subalgebra of $[0,\infty]^{[0,\infty]}$.

□

2.3.4 Theorem *There exists a one-to-one correspondence between the class $\mathfrak{S}$ of bireflective bicoreflective subconstructs of* **AP** *and the set $\mathfrak{M}$ of monoidal subalgebras of $[0,\infty]^{[0,\infty]}$ which are contained in $\Gamma_{\mathbf{TOP}}$. This correspondence is given by the bijection*

$$\mathbf{T} : \mathfrak{M} \longrightarrow \mathfrak{S} : \Gamma \longrightarrow \mathbf{T}(\Gamma),$$

the inverse of which is given by

$$\Omega : \mathfrak{S} \longrightarrow \mathfrak{M} : \mathcal{C} \longrightarrow \Omega(\mathcal{C}).$$

Proof If $\mathcal{C}$ is a bireflective bicoreflective subconstruct of **AP**, then it follows that $|\mathbf{TOP}| \subset |\mathcal{C}|$ since **TOP** itself has no such subconstructs. The remainder then follows at once from 2.3.2 and 2.3.3.

□

The task of finding all bireflective bicoreflective subconstructs of **AP** is now reduced to that of determining all monoidal subalgebras of $[0,\infty]^{[0,\infty]}$ which are contained in $\Gamma_{\mathbf{TOP}}$.

For any $m \in [0,\infty]$, we define the subset $\Gamma_m \subset [0,\infty]^{[0,\infty]}$ which consists of those functions $\varphi \in [0,\infty]^{[0,\infty]}$ which are left continuous and increasing and which fulfill

$$x < y \;\Rightarrow\; \varphi(y) - \varphi(x) \leq (y - x) \vee m.$$

For any set X and any $A \subset X$, we denote by $\mathbf{1}_A$ the characteristic function of A.

2.3.5 Lemma *For any $m \in [0,\infty]$, Γ_m is a monoidal subalgebra of $[0,\infty]^{[0,\infty]}$ such that $m\mathbf{1}_{]0,\infty]} \in \Gamma_m$ and $\Gamma_m \subset \Gamma_{\mathbf{TOP}}$.*

Proof Only the fact that Γ_m is closed under composition needs some verification. Let $\varphi \in \Gamma_m$ and $\psi \in \Gamma_m$. Then $\psi \circ \varphi$ is increasing and left continuous. If $x < y$, then either $\varphi(x) = \varphi(y)$ and it follow that $\psi \circ \varphi(x) = \psi \circ \varphi(y)$, or $\varphi(x) < \varphi(y)$ and it follows that $\psi \circ \varphi(y) - \psi \circ \varphi(x) \leq (\varphi(y) - \varphi(x)) \vee m \leq (y - x) \vee m$.

□

2.3.6 Lemma *If Γ is a monoidal subalgebra of $[0,\infty]^{[0,\infty]}$, different from $\Gamma_{\mathbf{AP}}$ and included in $\Gamma_{\mathbf{TOP}}$, then there exists an $m \in]0,\infty[$ such that $m\mathbf{1}_{]0,\infty]} \in \Gamma$.*

Proof Since $\Gamma_{\mathbf{AP}} \neq \Gamma$, there exist $\varphi \in \Gamma$ and $x, y \in [0,\infty]$ such that $x < y$ and such that $\varphi(y) - \varphi(x) > y - x$. Put $p \doteqdot \varphi(y) - \varphi(x)$ and $a \doteqdot y - x$. Since $\oplus_x \in \Gamma_{\mathbf{AP}}$ it follows that $\theta \doteqdot (\varphi \circ \oplus_x - \varphi(x)) \wedge p \in \Gamma$. Clearly, we have

$$\theta(0) = 0 \text{ and } \theta_{|[y-x,\infty]} = p. \tag{*}$$

Claim: if there exists $\theta \in \Gamma$ satisfying (*), then there exist $\tau \in \Gamma$ and $b \in]0,a[$ such that

$$\tau(0) = 0 \text{ and } \tau_{|]b,\infty]} = p.$$

Indeed, since $\theta \in \Gamma$ is left continuous, there exists $b \in]0,a[$ such that

$$b \leq t < a \Rightarrow a < \theta(t) < p.$$

Hence, if we put $\tau \doteqdot \theta \circ \theta$, then it follows that $\tau \in \Gamma$, $\tau(0) = \theta(0) = 0$, and, since $\theta(b) > a$, also that

$$t \geq b \Rightarrow \tau(t) = \theta \circ \theta(t) \geq \theta(\theta(b)) = p.$$

Now, let

$$\sigma \doteqdot \sup \left\{\varphi \in \Gamma \mid \varphi(0) = 0,\ \exists z \in]0,\infty[\ :\ \varphi_{|[z,\infty]} = p\right\}$$

and let

$$c \doteqdot \inf \{t \mid \sigma(t) = p\}.$$

The above claim shows that $c > 0$ and $\sigma(c) = p$ is impossible. Two other cases can now occur. Either $c = 0$, in which case $\sigma = p\mathbf{1}_{]0,\infty]}$ and we can take $m \doteqdot p$, or $c > 0$ and $p > \sigma(c)$ and then, since $\sigma(c+) = p$ and since $\sigma \circ \oplus_c - \sigma(c) \in \Gamma$, we can take $m \doteqdot p - \sigma(c)$.

□

2.3.7 Lemma *If Γ is a monoidal subalgebra of $[0,\infty]^{[0,\infty]}$ such that $m\mathbf{1}_{]0,\infty]} \in \Gamma$, for some $m \in]0,\infty[$, then Γ contains each of the functions*

$$\varphi_{p,q,r} \doteqdot p \vee (p+q)\mathbf{1}_{]r,\infty]},$$

$$\rho_{p,q,r,b} \doteqdot \varphi_{p,q,r} \wedge \oplus_b,$$

and

$$\sigma_{p,q,r,b} \doteqdot \varphi_{p,q,r} \wedge \ominus_b,$$

for all $p \in [0,\infty]$, $q \in [0,m]$, $r \in [0,\infty[$, and $b \in [0,\infty[$.

Proof Since $\ominus_r \in \Gamma_{\mathbf{AP}}$ the result for the first function in the above list follows from the fact that $\varphi_{p,q,r} = (m\mathbf{1}_{]0,\infty]} \circ \ominus_r + p) \wedge (p+q)$. For the remaining functions the result then follows from the facts that $\oplus_b \in \Gamma_{\mathbf{AP}}$ and $\ominus_b \in \Gamma_{\mathbf{AP}}$. □

2.3.8 Lemma *If Γ is a monoidal subalgebra of $[0,\infty]^{[0,\infty]}$, if $m\mathbf{1}_{]0,\infty]} \in \Gamma$, for some $m \in]0,\infty[$, and if $\gamma \in \Gamma_m$, $\epsilon > 0$, and $x \in]0,\infty[$, then there exists a function $\theta_{x,\varepsilon} \in \Gamma$ such that $\theta_{x,\varepsilon} \leq \gamma$ and $\gamma(x) - \varepsilon \leq \theta_{x,\varepsilon}(x)$.*

Proof We use the notations of the foregoing lemma. It is clear that we can always suppose $0 < \epsilon < \gamma(x) \wedge m$. Since γ is left continuous, we can find $\delta_0 \in]0,x[$ such that

$$x - \delta_0 \leq t \leq x \;\Rightarrow\; \gamma(x) - \frac{\epsilon}{2} < \gamma(t) \leq \gamma(x).$$

We now have to distinguish between four cases.

Case 1. $\gamma(x) \geq x \geq m$. We take $\delta \in]0, \delta_0 \wedge \frac{\epsilon}{2}]$ and define

$$\theta_{x,\epsilon} \doteqdot \rho_{\gamma(x)-m,\, m-\frac{\epsilon}{2},\, x-\delta,\, \gamma(x)-x}.$$

We can then easily verify that $\theta_{x,\varepsilon} \leq \gamma$ by successively considering the intervals $[0, x-m]$, $]x-m, x-\delta]$, $]x-\delta, x]$, and $]x,\infty]$.

Case 2. $\gamma(x) \geq x$, $x \leq m$. We now choose $\delta \in]0, \delta_0 \wedge \frac{\epsilon}{2} \wedge x]$ and define

$$\theta_{x,\epsilon} \doteqdot \varphi_{(\gamma(x)-m)\vee 0,\, (\gamma(x)-\frac{\epsilon}{2})\wedge(m-\frac{\epsilon}{2}),\, x-\delta}.$$

In this case we verify that $\theta_{x,\epsilon} \leq \gamma$ by considering the intervals $[0, x-\delta]$, $]x-\delta, x]$, and $]x,\infty]$.

Case 3. $m \leq \gamma(x) < x$. We now take $\delta \in]0, \delta_0 \wedge \frac{\epsilon}{2}]$ and define

$$\theta_{x,\epsilon} \doteqdot \sigma_{\gamma(x)-m,m-\frac{\epsilon}{2},x-\delta,x-\gamma(x)}.$$

In order to verify that $\theta_{x,\epsilon} \leq \gamma$, we now have to consider the intervals $[0, x - \gamma(x)]$, $]x - \gamma(x), x - m]$, $]x - m, x - \delta]$, $]x - \delta, x]$, and $]x, \infty]$.

Case 4. $\gamma(x) < x$, $\gamma(x) < m$. For this last case we choose $\delta \in]0, \delta_0 \wedge \frac{\epsilon}{2} \wedge x]$ and define

$$\theta_{x,\epsilon} \doteqdot \varphi_{0,\gamma(x)-\frac{\epsilon}{2},x-\delta}.$$

Once more the inequality $\theta_{x,\epsilon} \leq \gamma$ is verified by considering the intervals $[0, x - \delta]$, $]x - \delta, x]$, and $]x, \infty]$.

Finally, in all four cases, it is immediately seen that $\theta_{x,\epsilon}(x) = \gamma(x) - \frac{\epsilon}{2} > \gamma(x) - \epsilon$.

□

2.3.9 Proposition *If Γ is a monoidal subalgebra of $[0, \infty]^{[0,\infty]}$ and if $m\mathbf{1}_{]0,\infty]} \in \Gamma$ for some $m \in]0, \infty[$, then $\Gamma_m \subset \Gamma$.*

Proof If $\gamma \in \Gamma_m$, then, making use of 2.3.8, for each $x \in]0, \infty[$ and each $\epsilon > 0$, we can find $\theta_{x,\epsilon} \in \Gamma$ such that $\theta_{x,\epsilon} \leq \gamma$ and $\gamma(x) - \epsilon < \theta_{x,\epsilon}(x)$. It then follows that

$$\gamma = \gamma(0) \vee \sup_{x \in]0,\infty[} \sup_{\epsilon > 0} \theta_{x,\epsilon}$$

and thus $\gamma \in \Gamma$.

□

2.3.10 Corollary *For any $m \in]0, \infty[$, we have that $\{m\mathbf{1}_{]0,\infty]}\}^{\sim} = \Gamma_m \subset \Gamma_{\mathbf{TOP}}$.*

□

2.3.11 Theorem *1. The only monoidal subalgebras of $[0, \infty]^{[0,\infty]}$ which are included in $\Gamma_{\mathbf{TOP}}$ are the subalgebras Γ_m for $m \in [0, \infty]$.*

2. The only bireflective bicoreflective subconstructs of **AP** *are the subconstructs $\mathbf{AP}_m \doteqdot \mathbf{T}(\Gamma_m)$ for $m \in [0, \infty]$, with $\mathbf{AP}_0 = \mathbf{AP}$ and $\mathbf{AP}_\infty = \mathbf{TOP}$. They form a set which is linearly ordered by inclusion:*

$$p > q \Rightarrow |\mathbf{AP}_q| \subset |\mathbf{AP}_p|.$$

Proof Since each Γ_m is a monoidal subalgebra of $[0,\infty]^{[0,\infty]}$ satisfying $\Gamma_m \subset \Gamma_{\mathbf{TOP}}$ and since clearly $m\mathbf{1}_{]0,\infty]} \in \Gamma_p$ if and only if $m \leq p$, the following definition makes sense for any Γ such that $\Gamma \subset \Gamma_{\mathbf{TOP}}$:

$$m_\Gamma \doteqdot \sup\left\{p \in [0,\infty] \mid p\mathbf{1}_{]0,\infty]} \in \Gamma\right\}.$$

Note, moreover, that since $m\mathbf{1}_{]0,\infty]} = \sup_{p<m} p\mathbf{1}_{]0,\infty]}$, if $m_\Gamma \neq \infty$, the supremum is actually a maximum. Furthermore it is easily seen that $\Gamma_0 = \Gamma_{\mathbf{AP}}$ and $\Gamma_\infty = \Gamma_{\mathbf{TOP}}$. The result now follows at once from the foregoing results and from the observation that $\Gamma_{m_\Gamma} = \Gamma$.

□

We will now give internal characterizations of the objects in $\mathbf{AP}_m \doteqdot \mathbf{T}(\Gamma_m)$ in terms of various approach structures. We will restrict our attention to those structures which allow for an elegant and illuminating characterization, namely distance, tower, limit operator, approach system, and, of course, regular function frame.

2.3.12 Proposition *For an approach space with support X and for $m \in [0,\infty]$ the following properties of the respective approach structures are equivalent:*

1. $\delta(X \times 2^X) \subset \{0\} \cup [m,\infty]$,
2. $\forall \epsilon \in [0,m[\ : \ \mathfrak{t}_\epsilon = \mathfrak{t}_0$,
3. $\{\lambda\mathcal{F}(x) \mid \mathcal{F} \in \mathbf{F}(X),\ x \in X\} \subset \{0\} \cup [m,\infty]$,
4. $\forall x \in X$, $\forall \varphi \in \mathcal{A}(x)$, $\forall \epsilon \in]0,m[\ : \ \varphi \vee m\mathbf{1}_{\{\varphi \in]\epsilon,m[\}} \in \mathcal{A}(x)$,
5. $\forall \mu \in \mathcal{R}$ such that $\{\mu = 0\} \neq \emptyset \ : \ m\mathbf{1}_{\{\mu>0\}} \in \mathcal{R}$.

Proof The cases $m = 0$ and $m = \infty$, i.e. **AP** and **TOP** respectively, are trivial. Hence we may suppose that $m \in]0,\infty[$.

1 ⇒ 2. This follows immediately from 1.8.19.

2 ⇒ 3. This follows from the following formula which appears in the tables preceding 1.9,

$$\lambda\mathcal{F}(x) = \sup_{A \in \sec(\mathcal{F})} \inf\left\{\epsilon \in \mathbb{R}^+ \mid x \in \mathfrak{t}_\epsilon(A)\right\} \qquad \forall x \in X.$$

3 ⇒ 4. Let $x \in X$ and $\varphi \in \mathcal{A}(x)$. Then, for any ultrafilter $\mathcal{U}$ on X, we have that $\sup_{U \in \mathcal{U}} \inf_{y \in U} \varphi(y) \leq \lambda\mathcal{U}(x)$. Again, this is a formula which appears in the tables

preceding 1.9. If $\epsilon \in]0, m[$ and we put $\varphi^* \doteqdot \varphi \vee m\mathbf{1}_{\{\varphi \in]\epsilon, m[\}}$, then we have to prove the analogous formula for φ^*.

Two cases can occur. If $\lambda\mathcal{U}(x) \geq m$, then there is nothing to prove since the change from φ to φ^* only involves values smaller than m. If $\lambda\mathcal{U}(x) = 0$, then it follows from 1.8.36 and 1.8.39 that $\sup_{U\in\mathcal{U}} \inf_{y\in U} \varphi(y) = 0$. Hence, for all $U \in \mathcal{U}$ and $\eta \in]0, \epsilon[$, there exists $y \in U$ such that $\varphi^*(y) = \varphi(y) < \eta$ and thus we also have that $\sup_{U\in\mathcal{U}} \inf_{y\in U} \varphi^*(y) = 0$.

4 $\Rightarrow$ 5. Let $\mu \in \mathcal{R}$ be such that $\{\mu = 0\} \neq \emptyset$ and put $\mu^* \doteqdot m\mathbf{1}_{\{\mu>0\}}$. If $x \in X$ is such that $\mu^*(x) \neq 0$, then $\mu^*(x) = m$ and $\mu(x) > 0$. Now, if we put $q \doteqdot (m \wedge \mu(x))/2$, then it follows from 1.8.22 and 1.8.28 that there exists a $\varphi \in \mathcal{A}(x)$ such that $\inf_{y\in X} (\mu + \varphi)(y) > q$ and thus $\{\mu = 0\} \subset \{\varphi > q\}$. If we now put $\varphi^* \doteqdot \varphi \vee m\mathbf{1}_{\{\varphi \in]q, m[\}}$, then we have that $\varphi^* \in \mathcal{A}(x)$ and since clearly $\inf_{y\in\{\mu=0\}} \varphi^*(y) \geq m$, we already have that $\inf_{y\in\{\mu=0\}} (\mu^* + \varphi^*)(y) \geq m$. However, since $\mu^*(y) = m$ if $\mu(y) > 0$, it follows that $\inf_{y\in X} (\mu^* + \varphi^*)(y) \geq m$ and therefore

$$\sup_{\varphi\in\mathcal{A}(x)} \inf_{y\in X} (\mu^* + \varphi)(y) \geq m = \mu^*(x),$$

which, from the arbitrariness of $x \in X$, 1.8.22, and 1.8.28 proves that $\mu^* \in \mathcal{R}$.

5 $\Rightarrow$ 1. If $A \subset X$ and $x \in X$ are such that $\delta(x, A) > 0$, then it follows from the fact that $\delta(x, A) = \sup \{\mu(x) \mid \mu \in \mathcal{R},\ \mu_{|A} = 0\}$ (again, see the tables preceding 1.9) that there exists a $\mu \in \mathcal{R}$ such that $\mu(x) \neq 0$ and $\mu_{|A} = 0$. Taking $\mu^* \doteqdot m\mathbf{1}_{\{\mu>0\}}$, it follows that $\mu^*(x) = m$ and ${\mu^*}_{|A} = 0$ and we immediately obtain that $\delta(x, A) \geq m$.

□

2.3.13 Theorem *For any $m \in [0, \infty]$ we have that $\mathbf{AP}_m \doteqdot \mathbf{T}(\Gamma_m)$ is the full subconstruct of* **AP** *with objects those approach spaces which satisfy one, and hence all, of the equivalent conditions of the foregoing proposition.*

Proof This follows from 2.3.12 and the fact that $\mathbf{T}(\Gamma_m) = \mathbf{T}(\{m\mathbf{1}_{]0,\infty]}\})$.

□

3
METRIC APPROACH SPACES

In the previous chapter we have seen that the description of topological spaces by means of the various equivalent approach structures did not yield "numerical" information. Of course this was to be expected, since the only numerical values which come into play in the setting of topological spaces are 0 and ∞. The situation already becomes considerably more interesting for metric spaces. What we discover in this chapter are the first indications that the "approach setup" allows for a canonical quantification of certain topological concepts. In particular in the setting of metric spaces, some of these quantifications with respect to convergence already exist in the literature on approximation theory. For the basic concepts and terminology concerning metric spaces we refer to Appendix B.

3.1 METRIC APPROACH STRUCTURES

Given an ∞pq-metric space (X, d), we associate with it a natural approach space in the following way. We define the function

$$\delta_d : X \times 2^X \longrightarrow [0, \infty] : (x, A) \longrightarrow \inf_{a \in A} d(x, a).$$

3.1.1 Proposition *Suppose* $(X, d) \in |pq\mathbf{MET}^\infty|$. *Then the function*

$$\delta_d : X \times 2^X \longrightarrow [0, \infty]$$

is a distance on X.

Proof This is merely the special case of 1.8.5 where we take for the gauge basis $\mathcal{G} \doteqdot \{d\}$.

□

The above expression for δ_d is of course well known and notationally often no distinction is made between d and δ_d. We emphasize, however, that for our purposes it is important to use different notations for an ∞pq-metric and for the distance derived from it in the sense of the foregoing definition. In the first place, the two functions have different domains and in the second place, they determine categorically different structures. We have seen that **TOP** is simultaneously bireflectively and bicoreflectively embedded in **AP**, and hence for topological approach spaces, it makes no difference whether we perform constructions, such as the making of limits, colimits, initial, and final structures, in **TOP** or in **AP**. However, for ∞pq-metric approach spaces, as we will see later in this chapter when we study the precise way in which $pq\mathbf{MET}^{\infty}$ is embedded in **AP**, it does make a difference whether we make initial structures of ∞pq-metric approach spaces in $pq\mathbf{MET}^{\infty}$ or in **AP**. Hence it is important to make the distinction.

3.1.2 Definition An approach space of type (X, δ_d), for some ∞pq-metric d on X, will be called a *metric approach space*. A distance of type δ_d will be called a *metric distance*.

As we have done for topological approach spaces, we will now describe the different approach structures associated with a metric distance. We will see that, unlike for topological spaces, where we basically rediscovered classical topological concepts, for ∞pq-metric spaces we obtain several new concepts.

In what follows, if $(x_n)_n$ is a sequence in X, we will denote by $\langle (x_n)_n \rangle$ the Fréchet filter generated by it. This means that

$$\langle (x_n)_n \rangle \doteqdot \{F \subset X \mid \exists n : \ \{x_m \mid m \geq n\} \subset F\}.$$

3.1.3 Proposition *Suppose $(X,d) \in |pq\mathbf{MET}^{\infty}|$. Then the adherence and limit operators in (X, δ_d) are determined by:*

1. $\alpha_d \mathcal{F}(x) = \sup_{F\in\mathcal{F}} \inf_{y\in F} d(x,y) \qquad \forall x \in X,\ \forall \mathcal{F} \in \mathbf{F}(X),$
2. $\lambda_d \mathcal{F}(x) = \inf_{F\in\mathcal{F}} \sup_{y\in F} d(x,y) \qquad \forall x \in X,\ \forall \mathcal{F} \in \mathbf{F}(X).$

Proof The first property is an immediate consequence of the definition of the adherence operator given in 1.8.34 and of the definition of δ_d. The second property follows from 1.8.30 and 3.1.5.

□

3.1.4 Corollary *Suppose $(X,d) \in |pq\mathbf{MET}^{\infty}|$. If $(x_n)_n$ is a sequence in X and $x \in X$, then we have:*

1. $\alpha_d\langle(x_n)_n\rangle(x) = \liminf_{n\to\infty} d(x, x_n)$,
2. $\lambda_d\langle(x_n)_n\rangle(x) = \limsup_{n\to\infty} d(x, x_n)$.

□

A limit operator of type λ_d will be called a *metric limit operator*. We will comment on the result of 3.1.4 in the following section.

Approach systems and gauges in pq**MET**$^\infty$ have very simple descriptions, as the following results show.

3.1.5 Proposition *Suppose* $(X, d) \in |pq\mathbf{MET}^\infty|$. *Then the approach system in* (X, δ_d) *is given by*

$$\mathcal{A}_d(x) \doteqdot \left\{\varphi \in [0,\infty]^X \mid \varphi \leq d(x,\cdot)\right\} \qquad \forall x \in X$$

and a basis for this system is given by

$$\mathcal{B}_d(x) \doteqdot \{d(x,\cdot)\} \qquad \forall x \in X.$$

Proof By 1.8.23 we have that, for all $x \in X$,

$$\mathcal{A}_d(x) = \left\{\varphi \in [0,\infty]^X \mid \forall A \in 2^X : \inf_{a\in A} \varphi(a) \leq \inf_{a\in A} d(x,a)\right\}.$$

Clearly, $\inf_{a\in A} \varphi(a) \leq \inf_{a\in A} d(x,a)$ holds, for all $A \subset X$, if and only if $\varphi(a) \leq d(x,a)$ holds, for all $a \in X$.

□

The above result clarifies our terminology of local distances. In a metric approach space the basic local distance at a point x is simply the ∞pq-metric "localized" at x. An approach system of type $\mathcal{A}_d$ will be called a *metric approach system*.

Gauges in pq**MET**$^\infty$ are even simpler than approach systems.

3.1.6 Proposition *Suppose* $(X, d) \in |pq\mathbf{MET}^\infty|$. *Then the gauge in* (X, δ_d) *is given by*

$$\mathcal{G}_d \doteqdot \{e \in pq\mathbf{M}^\infty(X) \mid e \leq d\}$$

and a basis for this gauge is given by $\mathcal{H} \doteqdot \{d\}$.

Proof Referring to 1.8.3 instead of to 1.8.23, this is precisely the same as the proof of the foregoing result.

□

3.1.7 Proposition *Suppose $(X, d) \in |pq\mathbf{MET}^\infty|$. Then the tower in (X, δ_d) is given by the family $(\mathfrak{t}^d_\epsilon)_{\epsilon\in\mathbb{R}^+}$ where*

$$\mathfrak{t}^d_\epsilon : 2^X \longrightarrow 2^X : A \longrightarrow \{x \in X \mid \delta_d(x, A) \leq \epsilon\} \qquad \forall\epsilon \in \mathbb{R}^+.$$

Proof This follows immediately from 1.8.19.

□

A tower of type $\mathfrak{t}^d$ will be called a *metric tower*. The pretopological closures $\mathfrak{t}^d_\epsilon$, $\epsilon \in \mathbb{R}^+$, are referred to, in Beer and Luchetti [11], as (ϵ-)*enlargement operators*.

3.1.8 Proposition *Suppose $(X, d) \in |pq\mathbf{MET}^\infty|$. Then the hull operator in (X, δ_d) is given by*

$$h_d(\mu)(x) \doteqdot \inf_{y\in X} (\mu(y) + d(x, y)) \qquad \forall x \in X, \ \forall\mu \in [0, \infty]^X.$$

Proof This follows from 1.8.15 and 3.1.6.

□

A hull operator of type h_d will be called a *metric hull operator*.

3.1.9 Proposition *Suppose $(X, d) \in |pq\mathbf{MET}^\infty|$. Then the regular function frame in (X, δ_d) is given by*

$$\mathcal{R}_d \doteqdot \left\{\mu \in [0, \infty]^X \mid \forall x, y \in X : \ \mu(x) \leq \mu(y) + d(x, y)\right\}.$$

Proof This follows from 1.8.22 and 3.1.8.

□

A regular function frame of type $\mathcal{R}_d$ will be called a *metric regular function frame*.

3.1.10 Remarks Making use of the results of this section it is possible to formulate some of the transitions which we have seen in 1.8 in a nicer way.

1. The transition from distances to gauges (1.8.3). Making use of 3.1.1, if δ is a distance on X, it follows that the associated gauge can be written as

$$\mathcal{G} = \{d \in pq\mathrm{M}^{\infty}(X) \mid \delta_d \leq \delta\}.$$

2. The transition from gauges to distances (1.8.5). Making use of 3.1.1, if $\mathcal{G}$ is a gauge on X, it follows that the associated distance can be written as

$$\delta = \sup_{d \in \mathcal{G}} \delta_d.$$

3. The transition from gauges to hull operators (1.8.15). Making use of 3.1.8, if $\mathcal{G}$ is a gauge on X, it follows that the associated hull operator can be written as

$$h = \sup_{d \in \mathcal{G}} h_d.$$

4. The transition from gauges to limit operators (1.8.31). Making use of 3.1.3, if $\mathcal{G}$ is a gauge on X, it follows that the associated limit operator can be written as

$$\lambda = \sup_{d \in \mathcal{G}} \lambda_d.$$

In the sequel we will freely use these new ways of describing these various transitions, but, if necessary, nevertheless refer to the original theorems.

The foregoing results show that we can identify ∞pq-metric spaces with certain types of approach spaces. The following results give internal characterizations of both ∞pq-metric approach spaces and ∞p-metric approach spaces.

3.1.11 Theorem *For an approach space (X, δ), the following are equivalent:*

1. (X, δ) is ∞pq-metric,

2. for all $x \in X$ and $\mathcal{A} \subset 2^X$, we have $\delta(x, \cup\mathcal{A}) = \inf_{A \in \mathcal{A}} \delta(x, A)$,

3. for all $x \in X$ and $A \subset X$, we have $\delta(x, A) = \inf_{a \in A} \delta(x, \{a\})$.

Proof The implications $1 \Rightarrow 2 \Rightarrow 3$ are evident. To prove $3 \Rightarrow 1$, it suffices to notice that

$$X \times X \longrightarrow [0, \infty] : (x, y) \longrightarrow \delta(x, \{y\})$$

defines an ∞pq-metric on X and that the third property means precisely that δ is the distance derived from this ∞pq-metric according to 3.1.1.

□

3.1.12 Theorem *An approach space* (X, δ) *is* ∞p*-metric if and only if*

$$\inf_{a \in A} \delta(a, B) = \inf_{b \in B} \delta(b, A) \qquad \forall A, B \in 2^X.$$

Proof The only-if part is evident. To show the if part, it suffices to notice that symmetry of δ follows from letting A and B be singletons and that, further, for any $a \in X$ and $B \in 2^X$, if we put $A \doteqdot \{a\}$, then $\delta(a, B) = \inf_{b \in B} \delta(b, \{a\}) = \inf_{b \in B} \delta(a, \{b\})$ and thus, by 3.1.11, δ is ∞pq-metric.

□

3.2 CONVERGENCE IN METRIC APPROACH SPACES

3.2.1 Proposition *Suppose* $(X, d) \in |pq\mathbf{MET}^\infty|$. *If* $\mathcal{F} \in \mathbf{F}(X)$ *and* $x \in X$, *then the following hold:*

1. $\mathcal{F} \dashv x$ *in* $(X, \mathcal{T}_d)$ *if and only if* $\alpha_d \mathcal{F}(x) = 0$,
2. $\mathcal{F} \to x$ *in* $(X, \mathcal{T}_d)$ *if and only if* $\lambda_d \mathcal{F}(x) = 0$.

Proof This follows at once from 3.1.3.

□

3.2.2 Proposition *Suppose* $(X, d) \in |p\mathbf{MET}^\infty|$. *If* $\mathcal{F} \in \mathbf{F}(X)$ *and* $x \in X$, *then the following are equivalent:*

1. $\mathcal{F} \to x$ *in* $(X, \mathcal{T}_d)$,
2. $\lambda_d \mathcal{F} = d(\cdot, x)$.

Proof 1 ⇒ 2. From 1.8.40, taking into account the fact that, for any $y \in X$, we have $\delta_d(y, \{x\}) = d(y, x)$, it follows that we have

$$\begin{aligned}\lambda_d \mathcal{F}(y) &\leq d(y, x) + \lambda_d \mathcal{F}(x)\\ &= d(y, x),\end{aligned}$$

which proves one inequality.

To prove the other inequality, let $\epsilon > 0$ be fixed and, for each $F \in \mathcal{F}$, choose an arbitrary $z_F \in F \cap B(x, \epsilon)$. Then we obtain

$$\begin{aligned}\lambda_d \mathcal{F}(y) + \epsilon &\geq \inf_{F \in \mathcal{F}} d(y, z_F) + d(z_F, x)\\ &\geq d(y, x),\end{aligned}$$

which by the arbitrariness of ϵ proves that also $d(y, x) \leq \lambda_d \mathcal{F}(y)$.

2 $\Rightarrow$ 1. This follows from 3.2.1.

□

The foregoing result does not hold in $pq\mathbf{MET}^\infty$, as the following example shows.

3.2.3 Example Consider the real line $\mathbb{R}$ equipped with the ∞pq-metric

$$d(x, y) \doteq (x - y) \vee 0 \qquad \forall x, y \in \mathbb{R}.$$

Fix points $x, y \in \mathbb{R}$ such that $x < y$. The filter stacky converges to x in $\mathcal{T}_d$. However, from 3.1.3 we have

$$\begin{aligned} \lambda_d(\text{stack}y)(z) &= \inf_{F \in \text{stack}y} \sup_{z \in F} d(x, z) \\ &= d(z, y) \end{aligned}$$

and $d(z, y) \neq d(z, x)$ whenever $x < z$.

3.2.4 Examples We now return to the examples which we mentioned in the introduction.

1. Consider the real line $\mathbb{R}$ with the usual Euclidean metric d_E, let $\epsilon > 0$, and consider the sequences $(x_n)_n$ and $(y_n)_n$, where

$$x_n \doteq \begin{cases} \epsilon & n \text{ even}, \\ -\epsilon & n \text{ odd} \end{cases}$$

and

$$y_n \doteq \begin{cases} n & n \text{ even}, \\ -n & n \text{ odd}. \end{cases}$$

For the Euclidean topology, neither of these sequences converges. It is clear, however, that from the metric point of view the sequences behave very differently. Let us see if the limit and adherence operators in $(\mathbb{R}, \delta_{d_E})$ give us interesting and intuitively correct information concerning the behavior of these sequences and if they discover the difference between them. From 3.1.4 we obtain for the adherence of the sequence $(x_n)_n$ at any point $x \in \mathbb{R}$,

$$\begin{aligned} \alpha_{d_E}\langle (x_n)_n \rangle (x) &= \liminf_{n \to \infty} |x - x_n| \\ &= \begin{cases} |x - \epsilon| & x \geq 0, \\ |x + \epsilon| & x \leq 0, \end{cases} \end{aligned}$$

and for the limit of $(x_n)_n$ at any point $x \in \mathbb{R}$,

$$\lambda_{d_E}\langle (x_n)_n\rangle(x) = \limsup_{n\to\infty} |x - x_n| = |x| + \epsilon.$$

The smaller the value of the adherence operator (respectively the limit operator) the closer the point comes to being an adherence point (respectively a limit point). There are two points where $\alpha_{d_E}\langle (x_n)_n\rangle$ attains the (necessarily minimal) value 0, namely $x = \epsilon$ and $x = -\epsilon$. These two points are also the "real" adherence points of the sequence $(x_n)_n$ with respect to the Euclidean topology.

By contrast, $\lambda_{d_E}\langle (x_n)_n\rangle$ nowhere attains the value 0. This accords with the fact that the sequence $(x_n)_n$ does not converge with respect to the Euclidean topology. However, $\lambda_{d_E}\langle (x_n)_n\rangle(x)$ does attain a minimal value at the point $x = 0$, which intuitively seems indeed to be the point which best approximates a limit point.

The following figure gives the graphs of $\alpha_{d_E}\langle (x_n)_n\rangle$ (dotted line) and $\lambda_{d_E}\langle (x_n)_n\rangle$ (full line).

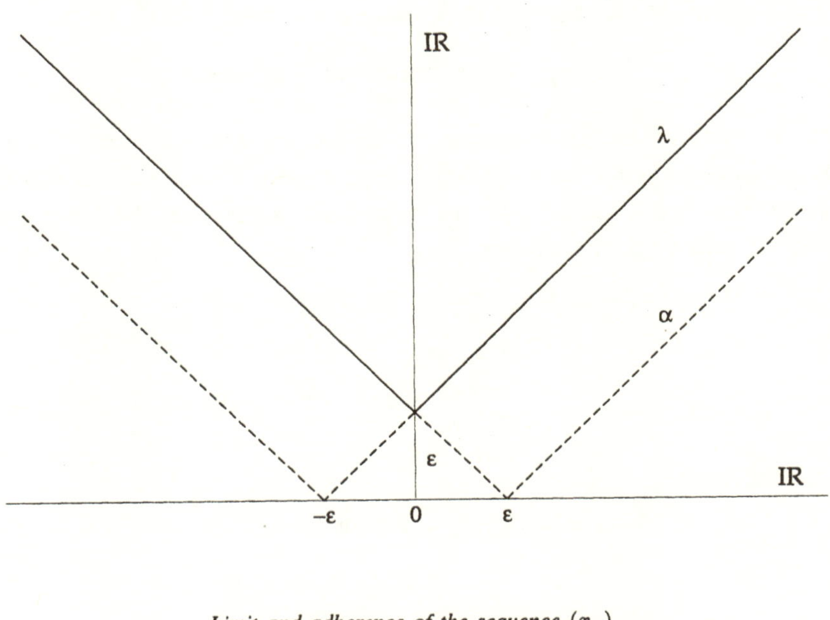

Limit and adherence of the sequence $(x_n)_n$

The figure suggests that $\alpha_{d_E}\langle(x_n)_n\rangle$ measures the distance to the nearest "real" adherence point, whereas $\lambda_{d_E}\langle(x_n)_n\rangle$ seems to measure the distance to the farthest "real" adherence point. Under certain conditions, fulfilled for the present example, this is indeed the case, as we will see in 3.2.8.

For the sequence $(y_n)_n$ we obtain that

$$\lambda_{d_E}\langle(y_n)_n\rangle(x) = \alpha_{d_E}\langle(y_n)_n\rangle(x) = \infty,$$

for all $x \in \mathbb{R}$. This sequence "really does neither adhere nor converge" to any point on the real line.

2. Again consider the real line $\mathbb{R}$ with the usual Euclidean metric d_E. Let $\varphi : \mathbb{R} \longrightarrow]-\epsilon, \epsilon[$ be an increasing homeomorphism and let $(r_n)_n$ be an enumeration of the rationals. Consider the sequences $(r_n)_n$ and $(\varphi(r_n))_n$.

From 3.1.4 it follows that, for all $x \in \mathbb{R}$, the adherence of the sequence is given by $(r_n)_n$,

$$\alpha_{d_E}\langle(r_n)_n\rangle(x) = \liminf_{n\to\infty} |r_n - x| = 0.$$

This, according to 3.2.1, is as it should be, since, for the usual topology, the sequence adheres to every point in $\mathbb{R}$.

For the limit of the sequence $(r_n)_n$, also at any point $x \in \mathbb{R}$, we find

$$\lambda_{d_E}\langle(r_n)_n\rangle(x) = \limsup_{n\to\infty} |r_n - x| = \infty,$$

which, in view of the chaotic behavior of the sequence (not only from the topological point of view but also from the metric point of view), seems plausible.

For the sequence $(\varphi(r_n))_n$ on the other hand, again from 3.1.4 and for all $x \in \mathbb{R}$, we obtain for the adherence

$$\begin{aligned}\alpha_{d_E}\langle(\varphi(r_n))_n\rangle(x) &= \liminf_{n\to\infty} |\varphi(r_n) - x| \\ &= \delta_{d_E}(x,]-\epsilon, \epsilon[) \\ &= \begin{cases} |x| - \epsilon & x \le -\epsilon \text{ or } \epsilon \le x, \\ 0 & -\epsilon \le x \le \epsilon, \end{cases}\end{aligned}$$

and for the limit

$$\begin{aligned}\lambda_{d_E}\langle(\varphi(r_n))_n\rangle(x) &= \limsup_{n\to\infty} |\varphi(r_n) - x| \\ &= d_E(x, 0) + \epsilon \\ &= |x| + \epsilon.\end{aligned}$$

These results seem to fit completely with our (metric) intuition. Although, topologically, the behavior of the sequence $(\varphi(r_n))_n$ is the same as that of the sequence $(r_n)_n$, metrically it behaves more nicely. From the point of view of approximation theory, for a sufficiently small ϵ, $(\varphi(r_n))_n$ might be considered to be "almost convergent". The following figure gives the graphs of the functions $\alpha_{d_E}\langle(\varphi(r_n))_n\rangle$ (dotted line) and $\lambda_{d_E}\langle(\varphi(r_n))_n\rangle$ (full line).

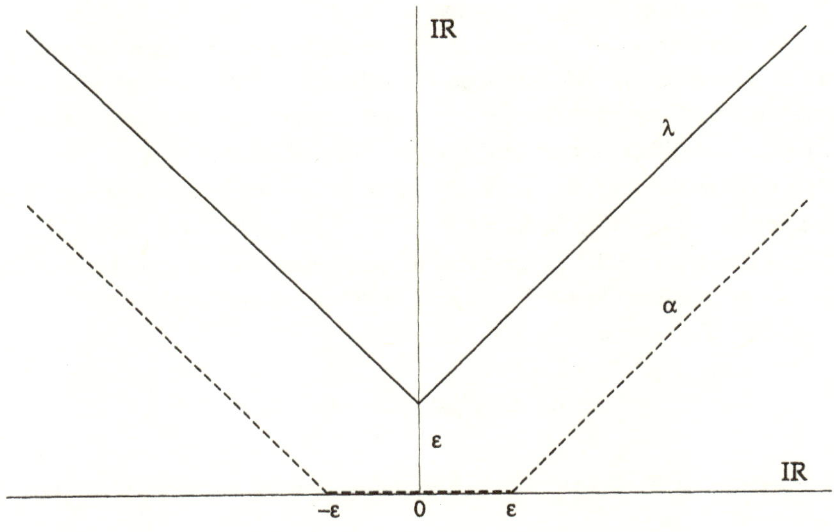

Limit and adherence of the sequence $(\varphi(r_n))_n$

3. Actually the ideas of the foregoing two items are not new. The formula given in 3.1.4 for the limit of a given sequence at a given point is well known in approximation theory. It was introduced in this field in 1972 by Edelstein in [32] and was later (1980) generalized for nets by Lim [65]. The setting there was mainly restricted to bounded sequences or nets in closed convex subsets of a Banach space and the main interest was to find a point where the "limit function" would be minimal. Such a point is called an *asymptotic center* and the value of the limit function at such an asymptotic center is called the *asymptotic radius*. Since an asymptotic center, if it exists, need not be unique, somewhat abusively,

the term asymptotic center is also used for the set of all asymptotic centers in the sense of the first definition. Some noteworthy properties, shown in that context, are the following:

1. in a uniformly convex Banach space a unique asymptotic center always exists,

2. in a weakly compact convex subset of a Banach space the asymptotic center is nonempty, closed, bounded, and convex,

3. for any filter $\mathcal{F}$, $\lambda_d\mathcal{F}$ is nonexpansive,

4. for any filter $\mathcal{F}$, $\lambda_d\mathcal{F}$ is convex,

5. in a Hilbert space or in a two-dimensional Banach space the asymptotic center of a bounded filter $\mathcal{F}$ intersects the adherence of the filter generated by the closed convex hulls of the members of $\mathcal{F}$ and, actually, this property characterizes the class of all Hilbert spaces and two-dimensional Banach spaces.

We refer to the literature for many other interesting results; see e.g. Garkavi [41], Edelstein [32], [33], Anderson *et al.* [2], Lim [65], and Holmes [52].

The examples in 3.2.4 suggest that in metric approach spaces the behavior of the limit operator is influenced by a sort of "Cauchyness" of the filter. We make this precise. For any set $F \subset X$, we put

$$\mathrm{diam}_d(F) \doteqdot \sup_{x,y\in F} d(x,y),$$

the *d-diameter* of F. Given a filter $\mathcal{F}$ in the ∞pq-metric space (X, d), we define the *width* of $\mathcal{F}$ as

$$\omega_d\mathcal{F} \doteqdot \inf_{F\in\mathcal{F}} \mathrm{diam}_d(F).$$

The width of $\mathcal{F}$ measures if $\mathcal{F}$ is a Cauchy filter, not with respect to the given space (X, d), but rather with respect to the associated ∞p-metric space (X, d^*) (see also Appendix B), where

$$d^*(x,y) \doteqdot d(x,y) \vee d(y,x) \qquad \forall x, y \in X.$$

Clearly then, $\omega_d\mathcal{F} = 0$ if and only if $\mathcal{F}$ is a Cauchy filter in (X, d^*).

3.2.5 Proposition *Suppose* $(X, d) \in |pq\mathbf{MET}^\infty|$. *If* $\mathcal{F} \in \mathbf{F}(X)$, *then we have*

$$\lambda_d\mathcal{F} \leq \alpha_d\mathcal{F} + \omega_d\mathcal{F}.$$

Proof If $\omega_d\mathcal{F} = \infty$ there is nothing to show. Suppose therefore that $\omega_d\mathcal{F} < \infty$, let $x \in X$, and let $\epsilon > 0$. Choose $F_\epsilon \in \mathcal{F}$ such that

$$\mathrm{diam}_d(F_\epsilon) \leq \omega_d\mathcal{F} + \epsilon.$$

For all $y, z \in F_\epsilon$ we then have that

$$d(x, y) \leq d(x, z) + \omega_d\mathcal{F} + \epsilon$$

and thus

$$\sup_{y \in F_\epsilon} d(x, y) \leq \delta_d(x, F_\epsilon) + \omega_d\mathcal{F} + \epsilon,$$

which implies that

$$\begin{aligned} \lambda_d\mathcal{F}(x) &= \inf_{F \in \mathcal{F}} \sup_{y \in F} d(x, y) \\ &\leq \sup_{y \in F_\epsilon} d(x, y) \\ &\leq \delta_d(x, F_\epsilon) + \omega_d\mathcal{F} + \epsilon \\ &\leq \alpha_d\mathcal{F} + \omega_d\mathcal{F} + \epsilon, \end{aligned}$$

which, by the arbitrariness of ϵ, proves the result.

□

3.2.6 Example The inequality in the foregoing proposition is best possible, as example 1 of the sequence $(x_n)_n$ in 3.2.4 shows. There we have that $\lambda_{d_E}\langle(x_n)_n\rangle(\epsilon) = \omega_{d_E}\langle(x_n)_n\rangle = 2\epsilon$ and $\alpha_{d_E}\langle(x_n)_n\rangle(\epsilon) = 0$. The fact that $\lambda_{d_E}\langle(x_n)_n\rangle(0) = \alpha_{d_E}\langle(x_n)_n\rangle(0) = \epsilon$, moreover, shows that, in general, the inequality can be strict.

3.2.7 Corollary *Suppose* $(X, d) \in |pq\mathbf{MET}^\infty|$. *If* $\mathcal{F} \in \mathbf{F}(X)$ *is a Cauchy filter in* (X, d^*) *then* $\lambda_d\mathcal{F} = \alpha_d\mathcal{F}$.

□

We recall that a filter is called *total* if all finer ultrafilters are convergent (not necessarily to the same point). This notion was introduced by Pettis in [99] and extensively used and studied by Vaughan in [115], [116]. In regular spaces, a filter $\mathcal{F}$ is total if and only if it is so-called *adherent–convergent*. This means that every open set containing the set of adherence points of $\mathcal{F}$ is a member of $\mathcal{F}$.

3.2.8 Proposition *Suppose* $(X, d) \in |\mathbf{pMET}^\infty|$. *If* $\mathcal{F} \in \mathbf{F}(X)$ *is total and* $x \in X$, *then the following hold:*

1. $\alpha_d\mathcal{F}(x) = \inf_{y \in \mathrm{adh}_{\mathcal{T}_d}\mathcal{F}} d(x, y) = \delta_d(x, \mathrm{adh}_{\mathcal{T}_d}\mathcal{F})$,
2. $\lambda_d\mathcal{F}(x) = \sup_{y \in \mathrm{adh}_{\mathcal{T}_d}\mathcal{F}} d(x, y)$.

Proof To prove the first property, suppose that $\delta_d(x, \mathrm{adh}_{\mathcal{T}_d}\mathcal{F}) < \epsilon$. Then there exists $y \in \bigcap_{F \in \mathcal{F}} \mathrm{cl}_{\mathcal{T}_d} F$ for which $d(x, y) < \epsilon$ and it follows that

$$\begin{aligned}\alpha_d\mathcal{F}(x) &= \sup_{F \in \mathcal{F}} \delta_d(x, F) \\ &= \sup_{F \in \mathcal{F}} \delta_d(x, \mathrm{cl}_{\mathcal{T}_d} F) \\ &\leq d(x, y) \\ &< \epsilon,\end{aligned}$$

which proves that $\alpha_d\mathcal{F}(x) \leq \delta_d(x, \mathrm{adh}_{\mathcal{T}_d}\mathcal{F})$. To prove the other inequality let $\epsilon > 0$. Since

$$(\mathrm{adh}_{\mathcal{T}_d}\mathcal{F})^{(\epsilon)} \supset \left\{\delta_{\mathrm{adh}_{\mathcal{T}_d}\mathcal{F}} < \epsilon\right\}$$

it follows from 2.2.10 and the fact that $\mathcal{F}$ is adherent–convergent (Pettis [99]) that $(\mathrm{adh}_{\mathcal{T}_d}\mathcal{F})^{(\epsilon)} \in \mathcal{F}$. Consequently

$$\begin{aligned}\delta_d(x, \mathrm{adh}_{\mathcal{T}_d}\mathcal{F}) &\leq \delta_d\left(x, (\mathrm{adh}_{\mathcal{T}_d}\mathcal{F})^{(\epsilon)}\right) + \epsilon \\ &\leq \alpha_d\mathcal{F}(x) + \epsilon.\end{aligned}$$

For the proof of the second property, since an ultrafilter finer than $\mathcal{F}$ converges to a point in $\mathrm{adh}_{\mathcal{T}_d}\mathcal{F}$, it follows from the first property, 1.8.36, and 1.8.37 that

$$\begin{aligned}\lambda_d\mathcal{F}(x) &= \sup_{\mathcal{U} \in \mathbf{U}(\mathcal{F})} \lambda_d\mathcal{U}(x) \\ &= \sup_{\mathcal{U} \in \mathbf{U}(\mathcal{F})} \delta_d(x, \mathrm{adh}_{\mathcal{T}_d}\mathcal{U}) \\ &\leq \sup_{y \in \mathrm{adh}_{\mathcal{T}_d}\mathcal{F}} d(x, y).\end{aligned}$$

For the converse inequality it suffices to notice that, for any $y \in \mathrm{adh}_{\mathcal{T}_d}\mathcal{F}$, we have

$$\begin{aligned}\lambda_d\mathcal{F}(x) &= \inf_{F \in \mathcal{F}} \sup_{z \in F} d(x, z) \\ &= \inf_{F \in \mathcal{F}} \sup_{z \in \mathrm{cl}_{\mathcal{T}_d}(F)} d(x, z) \\ &\geq \inf_{F \in \mathcal{F}} d(x, y) \\ &= d(x, y),\end{aligned}$$

which proves that $\sup_{y \in \mathrm{adh}_{\mathcal{T}_d} \mathcal{F}} d(x,y) \leq \lambda_d \mathcal{F}(x)$.

□

Again it should be pointed out that the foregoing result does not hold in $pq\mathbf{MET}^{\infty}$.

3.2.9 Example Consider the set of real numbers $[0,\infty[$ equipped with the ∞pq-metric d_P (see also the remarks before 1.8.33)

$$d_P(x,y) \doteqdot (x-y) \vee 0 \qquad x,y \in [0,\infty[.$$

Fix a point $x_0 \in]0,\infty[$. For the filter $\mathcal{F} \doteqdot \mathrm{stack} x_0$ we have $\mathrm{adh}_{\mathcal{T}_{d_P}} \mathcal{F} = [0, x_0]$. Consequently it follows that, for all $x \in [0,\infty[$,

$$\sup_{y \in \mathrm{adh}_{\mathcal{T}_{d_P}} \mathcal{F}} d_P(x,y) = d_P(x,0),$$

whereas

$$\lambda_{d_P} \mathcal{F}(x) = d_P(x, x_0).$$

Totality too is a necessary condition in 3.2.8.

3.2.10 Example Consider $\mathbb{R} \setminus \{0\}$ equipped with the usual metric d_E and the usual topology $\mathcal{T}_E$, and consider the sequence $(z_n)_{n \geq 1}$, where

$$z_n \doteqdot \begin{cases} n & n \text{ even}, \\ \frac{1}{n} & n \text{ odd}. \end{cases}$$

Denote by $\mathcal{F}$ the filter generated by $(z_n)_{n \geq 1}$. Then the adherence and limit of $\mathcal{F}$ are given respectively by

$$\alpha_{d_E} \mathcal{F}(x) = |x| \qquad \forall x \in \mathbb{R} \setminus \{0\}$$

and

$$\lambda_{d_E} \mathcal{F}(x) = \infty \qquad \forall x \in \mathbb{R} \setminus \{0\}.$$

Now clearly, $\mathcal{F}$ is not total and, actually, $\mathrm{adh}_{\mathcal{T}_E} \mathcal{F} = \emptyset$. Consequently, for any $x \in \mathbb{R} \setminus \{0\}$, the formulas of 3.2.8 give us $\inf_{y \in \mathrm{adh}_{\mathcal{T}_E} \mathcal{F}} d(x,y) = \infty$, which differs from $\alpha_{d_E} \mathcal{F}(x)$ and $\sup_{y \in \mathrm{adh}_{\mathcal{T}_E} \mathcal{F}} d(x,y) = 0$, which differs from $\lambda_{d_E} \mathcal{F}(x)$.

3.3 THE EMBEDDING OF $pq\mathbf{MET}^\infty$ IN AP

In 3.1 we have seen that ∞pq-metric spaces can be viewed as special types of approach spaces. That $pq\mathbf{MET}^\infty$ is concretely embedded in **AP** is a consequence of the following proposition.

3.3.1 Proposition *Let (X, d) and (X', d') be ∞pq-metric spaces and consider the function $f : X \longrightarrow X'$. Then the following are equivalent:*

1. *$f : (X, d) \longrightarrow (X', d')$ is nonexpansive,*
2. *$f : (X, \delta_d) \longrightarrow (X', \delta_{d'})$ is a contraction.*

Proof This follows from 1.9.2, using any of the descriptions of the various equivalent approach structures in ∞pq-metric approach spaces given in 3.1.

□

3.3.2 Corollary *The functor*

$$\begin{array}{ccc} pq\mathbf{MET}^\infty & \rightarrow & \mathbf{AP} \\ (X,d) & \rightarrowtail & (X,\delta_d) \\ f & \rightarrowtail & f \end{array}$$

is a full concrete embedding of $pq\mathbf{MET}^\infty$ into **AP**.

□

In the second chapter we were able to show both bireflectivity and bicoreflectivity of the embedding of **TOP** in **AP**. For $p\mathbf{MET}^\infty$ and $pq\mathbf{MET}^\infty$ we are only able to show bicoreflectivity of the embedding. As we will see later, it is precisely the fact that neither $p\mathbf{MET}^\infty$ nor $pq\mathbf{MET}^\infty$ is embedded bireflectively in **AP** which makes the theory of approach spaces especially interesting.

3.3.3 Theorem *The construct $pq\mathbf{MET}^\infty$ of ∞pq-metric spaces and nonexpansive functions is embedded as a concretely bicoreflective subconstruct of* **AP**.

For any space $(X, \delta) \in |\mathbf{AP}|$, its $pq\mathbf{MET}^\infty$-bicoreflection is given by

$$id_X : (X, \delta^{qm}) \longrightarrow (X, \delta),$$

where δ^{qm} is the distance determined by the ∞pq-metric

$$d_\delta : X \times X \longrightarrow [0, \infty] : (x, y) \longrightarrow \delta(x, \{y\}).$$

Proof To show that $id_X : (X, \delta^{qm}) \longrightarrow (X, \delta)$ is a contraction let $x \in X$ and let $A \subset X$. Then we have

$$\begin{aligned}\delta(x, A) &\leq \inf_{a \in A} \delta(x, \{a\}) \\ &= \delta^{qm}(x, A).\end{aligned}$$

Now suppose that $(Y, d) \in |pq\mathbf{MET}^\infty|$ and that

$$f : (Y, \delta_d) \longrightarrow (X, \delta)$$

is a contraction. Then, for any $x \in Y$ and $A \subset Y$, we have

$$\begin{aligned}\delta^{qm}(f(x), f(A)) &= \inf_{a \in A} \delta(f(x), \{f(a)\}) \\ &\leq \inf_{a \in A} \delta_d(x, \{a\}) \\ &= \delta_d(x, A),\end{aligned}$$

which proves that

$$f : (Y, \delta_d) \longrightarrow (X, \delta^{qm})$$

is also a contraction.

□

3.3.4 Corollary *pq**MET**$^\infty$ is closed under the formation of colimits and final structures in* **AP**. *In particular, a coproduct in* **AP** *of a family of ∞pq-metric approach spaces is an ∞pq-metric approach space and, likewise, a quotient in* **AP** *of an ∞pq-metric approach space is an ∞pq-metric approach space.*

Proof This follows from 3.3.3 and Adámek *et al.* [1].

□

In the following result we describe the pq**MET**$^\infty$-bicoreflection of an approach space by means of some of the most important approach structures. This result is the counterpart of 2.2.8. Note that the ∞pq-metric generating the pq**MET**$^\infty$-bicoreflection of an approach space (X, δ) is denoted by d_δ (see 3.3.3).

3.3.5 Proposition *Let $(X, \delta) \in |\mathbf{AP}|$. Then the following hold:*

1. *if $(\mathcal{B}(x))_{x \in X}$ is a basis for the approach system then*

$$d_\delta(x, y) = \sup_{\varphi \in \mathcal{B}(x)} \varphi(y) \qquad \forall x, y \in X,$$

2. *if $\mathcal{H}$ is a basis for the gauge associated with δ then*

$$d_\delta(x, y) = \sup_{d \in \mathcal{H}} d(x, y) \qquad \forall x, y \in X.$$

Proof The first property follows from 1.8.24 and 3.3.3 while the second property follows from 1.8.5 and 3.3.3.

□

3.3.6 Theorem *The construct* $p\mathbf{MET}^\infty$ *of ∞p-metric spaces and nonexpansive functions is embedded as a concretely bicoreflective subconstruct of* $\mathbf{AP}$.

For any space $(X, \delta) \in |\mathbf{AP}|$, its $p\mathbf{MET}^\infty$-bicoreflection is given by

$$id_X : (X, \delta^m) \longrightarrow (X, \delta),$$

where δ^m is the distance determined by the ∞p-metric

$$d_\delta^* : X \times X \longrightarrow [0, \infty] : (x, y) \longrightarrow \delta(x, \{y\}) \vee \delta(y, \{x\}).$$

Proof This proof is precisely the same as the one from 3.3.3.

□

3.3.7 Corollary $p\mathbf{MET}^\infty$ *is closed under the formation of colimits and final structures in* $\mathbf{AP}$. *In particular, a coproduct in* $\mathbf{AP}$ *of a family of ∞p-metric approach spaces is an ∞p-metric approach space and, likewise, a quotient in* $\mathbf{AP}$ *of an ∞p-metric approach space is an ∞p-metric approach space.*

Proof This follows from 3.3.6 and Adámek *et al.* [1].

□

From 3.3.3 and 3.3.6 it follows that $d_\delta^* = (d_\delta)^*$ where, for any $d \in pq\mathrm{M}^\infty(X)$, d^* stands for the symmetrization of d, given by $d^*(x, y) \doteqdot d(x, y) \vee d(y, x)$, for all $x, y \in X$. The description of the $p\mathbf{MET}^\infty$-bicoreflection of an approach space by means of a base for the approach system or a base for the gauge is therefore easily deduced from 3.3.5 and 3.3.6. For instance, if $\mathcal{H}$ is a basis for the gauge associated with δ, then $d_\delta^*(x, y) = \sup_{d \in \mathcal{H}} d^*(x, y)$, for all $x, y \in X$.

3.3.8 Examples Again we refer to the examples which we considered in 1.8.33.

The distance of the first example is $\delta_{\mathbb{E}} : [0,\infty] \times 2^{[0,\infty]} \longrightarrow [0,\infty]$, where

$$\delta_{\mathbb{E}}(x, A) \doteqdot \begin{cases} 0 & x = \infty, A \text{ unbounded}, \\ \infty & x = \infty, A \text{ bounded}, \\ \inf\limits_{a \in A \cap \mathbb{R}^+} |x - a| & x < \infty. \end{cases}$$

Both the $pq\mathbf{MET}^\infty$-bicoreflection and the $p\mathbf{MET}^\infty$-bicoreflection of this space are given by $([0,\infty], d_{\mathbb{E}})$ where $d_{\mathbb{E}}$ is the "Euclidean" metric on $[0,\infty]$, i.e.

$$d_{\mathbb{E}}(x, y) \doteqdot |x - y| \qquad \forall x, y \in [0,\infty].$$

The distance of the second example is $\delta_{\mathbb{P}} : [0,\infty] \times 2^{[0,\infty]} \longrightarrow [0,\infty]$, where

$$\delta_{\mathbb{P}}(x, A) \doteqdot \begin{cases} (x - \sup A) \vee 0 & A \neq \emptyset, \\ \infty & A = \emptyset. \end{cases}$$

The $pq\mathbf{MET}^\infty$-bicoreflection of this space is $([0,\infty], d_{\mathbb{P}})$, where

$$d_{\mathbb{P}}(x, y) \doteqdot (x - y) \vee 0 \qquad \forall x, y \in [0,\infty]$$

and the $p\mathbf{MET}^\infty$-bicoreflection is given by $([0,\infty], d_{\mathbb{E}})$.

3.3.9 Proposition *Let (X, δ) be an approach space. Then, for any $\xi \in \mathcal{R}$,*

$$\xi : (X, d_\delta) \longrightarrow ([0,\infty], d_{\mathbb{P}})$$

is a nonexpansive map. In particular:

1. *for any $A \in 2^X$, the distance functional*

$$\delta_A : (X, d_\delta) \longrightarrow ([0,\infty], d_{\mathbb{P}})$$

is a nonexpansive map,

2. *for any $\mathcal{F} \in \mathbf{F}(X)$, both*

$$\lambda\mathcal{F} : (X, d_\delta) \longrightarrow ([0,\infty], d_{\mathbb{P}})$$

and

$$\alpha\mathcal{F} : (X, d_\delta) \longrightarrow ([0,\infty], d_{\mathbb{P}})$$

are nonexpansive maps.

Proof This follows at once from 1.9.6, 3.3.3, and 3.3.8.

□

From 3.1.9 it follows at once that $\xi \in \mathcal{R}_d$ if and only if the function $\xi : (X, d) \longrightarrow ([0, \infty], d_{\mathbb{P}})$ is nonexpansive, which means that for ∞pq-metric approach spaces the condition given in 3.3.9 is both necessary and sufficient. This also implies that, for any $\mu \in [0, \infty]^X$, the hull $h_d(\mu)$, as given in 3.1.8, can be described as the largest nonexpansive function smaller than μ. This alternative characterization is well known in the classical situation (i.e. when considering only real-valued functions), and can for instance be found in Singer [111].

In order to illustrate further the **TOP**-bicoreflection, the $pq\mathbf{MET}^{\infty}$-bicoreflection, and the $p\mathbf{MET}^{\infty}$-bicoreflection of an approach space, we will now describe these respectively for a topological, an ∞pq-metric, and an ∞p-metric approach space.

3.3.10 Proposition *If $(X, \delta_{\mathcal{T}})$ is a topological approach space, then:*

1. the $pq\mathbf{MET}^{\infty}$-bicoreflection is given by (X, δ_{d_1}), where d_1 is the ∞pq-metric

$$d_1(x,y) \doteqdot \begin{cases} 0 & x \in \mathrm{cl}_{\mathcal{T}}\{y\}, \\ \infty & x \notin \mathrm{cl}_{\mathcal{T}}\{y\}, \end{cases}$$

and

2. the $p\mathbf{MET}^{\infty}$-bicoreflection is given by (X, δ_{d_0}), where d_0 is the ∞p-metric

$$d_0(x,y) \doteqdot \begin{cases} 0 & x \in \mathrm{cl}_{\mathcal{T}}\{y\} \text{ and } y \in \mathrm{cl}_{\mathcal{T}}\{x\}, \\ \infty & x \notin \mathrm{cl}_{\mathcal{T}}\{y\} \text{ or } y \notin \mathrm{cl}_{\mathcal{T}}\{x\}. \end{cases}$$

Proof This follows from the definition of $\delta_{\mathcal{T}}$ and from 3.3.3 and 3.3.6.

□

The foregoing result has a rather interesting interpretation. In view of the fact that $x \notin \mathrm{cl}_{\mathcal{T}}\{y\}$ if and only if there exists a neighborhood V of x such that $y \notin V$, it follows that the $pq\mathbf{MET}^{\infty}$-bicoreflection indicates whether the topological space is T_1 or not. More precisely, $(X, \mathcal{T})$ is T_1 if and only if d_1 is an ∞-metric. In the same way the $p\mathbf{MET}^{\infty}$-bicoreflection indicates whether the topological space is T_0 or not. More precisely, $(X, \mathcal{T})$ is T_0 if and only if d_0 is an ∞-metric.

3.3.11 Proposition *If (X, δ_d) is an ∞pq-metric approach space, then:*

1. the **TOP**-*bicoreflection of (X, δ_d) is given by $(X, \delta_{\mathcal{T}_d})$, where $\mathcal{T}_d$ is the topology generated by d,*
and

2. the p**MET**$^\infty$-*bicoreflection of (X, δ_d) is given by (X, δ_{d^*}), where d^* is the ∞p-metric*

$$d^*(x, y) \doteqdot d(x, y) \vee d(y, x) \qquad \forall x, y \in X.$$

Proof The first property follows from the second property in 2.2.8. This result indeed implies that the **TOP**-bicoreflection of (X, δ_d) has as neighborhood system

$$\mathcal{V}(x) = \left\{V \in 2^X \mid \exists \epsilon > 0 : B_d(x, \epsilon) \subset V\right\} \qquad \forall x \in X.$$

The second property follows immediately from 3.3.6.

□

3.3.12 Proposition *If (X, δ_d) is an ∞p-metric approach space, then the* **TOP**-*bicoreflection of (X, δ_d) is given by $(X, \delta_{\mathcal{T}_d})$, where $\mathcal{T}_d$ is the topology generated by d.*

Proof This is exactly the same as the proof of 3.3.11.

□

The foregoing results, especially 3.3.11 and 3.3.12, are important for the sequel. A fundamental relationship among the different types of structures which we are considering in this work is that of a topology generated by a metric. As we argued in the introduction, it is the failure of this relation to be well behaved with respect to products which is one of the main motivations for considering approach spaces.

What the foregoing results tell us is that this relationship is recaptured in **AP** as a canonical functor, namely the **TOP**-bicoreflector restricted to $pq\mathbf{MET}^\infty$. In the case of an ∞pq-metric space the **TOP**-bicoreflector gives us the underlying topological space. It is natural therefore to extend this interpretation to the whole of **AP** and given an arbitrary approach space (X, δ), we will speak of (X, δ^{tc}) (the **TOP**-bicoreflection of (X, δ)) as the *underlying topological approach space* and of the topology generating (X, δ^{tc}) as the *topology underlying* δ or the topology *generated by* δ. This topology will be denoted by $\mathcal{T}_\delta$. The situation is clarified in the following diagram.

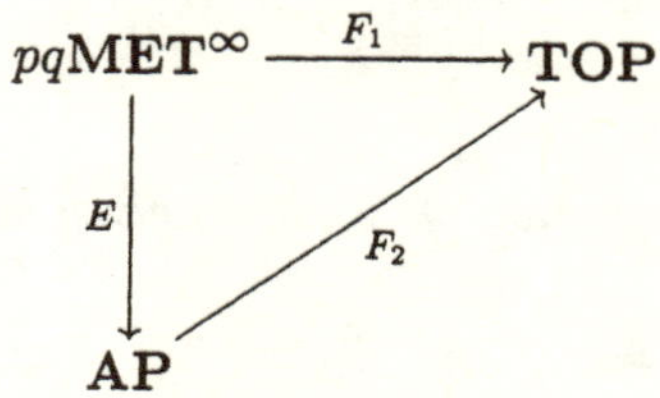

The functor E is the embedding of $pq\mathbf{MET}^\infty$ in $\mathbf{AP}$, F_1 is the forgetful functor associating with each ∞pq-metric space its underlying topological space, and F_2 is the $\mathbf{TOP}$-bicoreflector. The diagram commutes and F_2 thus is an extension of F_1.

It is a fundamental aspect of the theory of approach spaces that $pq\mathbf{MET}^\infty$ and, especially, $p\mathbf{MET}^\infty$ are not epireflectively embedded in $\mathbf{AP}$. In particular, neither subconstruct is stable for the formation of products in $\mathbf{AP}$. As the following result shows, this is the only reason why neither $pq\mathbf{MET}^\infty$ nor $p\mathbf{MET}^\infty$ are epireflectively embedded in $\mathbf{AP}$.

3.3.13 Proposition *$pq\mathbf{MET}^\infty$ and $p\mathbf{MET}^\infty$ are closed under the formation of subspaces in $\mathbf{AP}$.*

Proof This follows at once from the definitions.

□

There is still another noteworthy fact concerning the embedding of $\mathbf{TOP}$ and $pq\mathbf{MET}^\infty$ in $\mathbf{AP}$ which we would like to give. There are spaces which are at the same time topological and metric. The subconstructs to which they give rise are isomorphic to some remarkable subconstructs of $\mathbf{TOP}$. Let us denote the full subconstruct with objects all discrete spaces by $\mathbf{DIS}$, the subconstruct with objects all coproducts of indiscrete spaces by $\mathbf{CIN}$, and the subconstruct with objects all finitely generated spaces by $\mathbf{FING}$. We recall that a topological space is called finitely generated if a point is in the closure of a set A if and only if it is in the closure of one of the points of A. The abuse of notation in the following result is compensated for by the clarity of the formulation.

3.3.14 Proposition *The following properties hold:*

1. $\mathbf{TOP} \cap pq\mathbf{MET}^\infty = \mathbf{FING}$,
2. $\mathbf{TOP} \cap p\mathbf{MET}^\infty = \mathbf{CIN}$,

3. **TOP ∩ MET$^\infty$ = DIS.**

Proof 1. If $(X, \mathcal{T})$ is finitely generated, then, clearly, $\delta_\mathcal{T}$ fulfills the conditions of 3.1.11. Conversely, if (X, δ) is at the same time topological and ∞pq-metric, then it follows from 2.1.9 and 3.1.11 that, for any $x \in X$ and $A \subset X$, $\delta(x, A) = 0$ if and only if there exists $a \in A$ such that $\delta(x, \{a\}) = 0$. Hence the underlying topology is finitely generated.

2. If $(X, \mathcal{T})$ is a coproduct of indiscrete spaces, then it follows from the first property that $\delta_\mathcal{T}$ is at the same time topological and ∞pq-metric. Let $\delta_\mathcal{T} = \delta_d$ for some ∞pq-metric d. Then, if X is the coproduct of spaces $(X_j)_j$, it follows that, for any $x, y \in X$, $d(x, y) = 0$ if and only if there exists $j \in J$ such that $x, y \in X_j$. Hence d is symmetric. Conversely, if (X, δ) is at the same time topological and ∞p-metric, then it is easily seen that the open sets and the closed sets in the underlying topology coincide and hence, by Herrlich [47], this topology is a coproduct of indiscrete topologies.

3. If $(X, \mathcal{T})$ is discrete, then it follows from the second property that $\delta_\mathcal{T}$ is at the same time topological and ∞p-metric. The discreteness, however, implies that d is separated. Conversely, if (X, δ) is at the same time topological and ∞-metric, then it is easily seen that every subset of X coincides with its closure. □

3.4 THE EPIREFLECTIVE HULL OF pqMET$^\infty$

In this section we prove that **AP** is the right candidate to fill in the position of the second question mark in the diagram in the introduction. At the topological level we know that **TOP** consists precisely of subspaces of products (in **TOP**) of ∞pq-metrizable topological spaces, Herrlich [46]. We will show that this repeats itself at the "approach level".

In the following theorem we use the notation of 1.8.4. This means that, given an approach space (X, δ), $Z \subset X$, and $\zeta < \infty$, we consider the pq-metric d_Z^ζ which, for all $x, y \in X$, is given by

$$d_Z^\zeta(x, y) = (\delta(x, Z) \wedge \zeta - \delta(y, Z) \wedge \zeta) \vee 0.$$

3.4.1 Theorem *If $(X, \delta) \in |\mathbf{AP}|$ and we put $J \doteqdot \mathbb{R}^+ \times 2^X$, then we have that*

$$\psi : (X, \delta) \longrightarrow \left(X^J, \prod_{(\zeta, Z) \in J} \delta_{d_Z^\zeta} \right) : x \longrightarrow \left(x_{(\zeta, Z)} \doteqdot x \right)_{(\zeta, Z) \in J}$$

is an embedding.

Proof Let us put $\delta^* \doteqdot \prod_{(\zeta, Z) \in J} \delta_{d_Z^\zeta}$, the product distance on X^J. Let $x \in X$ and $A \subset X$. Then, making use of 1.10.4, on the one hand we have

$$\begin{aligned}
\delta^*(\psi(x), \psi(A)) = & \sup_{\mathcal{Z} \in 2^{(2^X)}} \sup_{\zeta \in \mathbb{R}^+} \inf_{a \in A} \sup_{Z \in \mathcal{Z}} d_Z^\zeta(x, a) \\
= & \sup_{\mathcal{Z} \in 2^{(2^X)}} \sup_{\zeta \in \mathbb{R}^+} \sup_{\varphi \in \mathcal{Z}^A} \inf_{a \in A} d_{\varphi(a)}^\zeta(x, a) \\
= & \sup_{\mathcal{Z} \in 2^{(2^X)}} \sup_{\zeta \in \mathbb{R}^+} \sup_{\varphi \in \mathcal{Z}^A} \inf_{Z \in \mathcal{Z}} \inf_{a \in \varphi^{-1}(Z)} d_Z^\zeta(x, a) \\
= & \sup_{\mathcal{Z} \in 2^{(2^X)}} \sup_{\zeta \in \mathbb{R}^+} \sup_{\varphi \in \mathcal{Z}^A} \inf_{Z \in \mathcal{Z}} \left(\delta(x, Z) \wedge \zeta - \sup_{a \in \varphi^{-1}(Z)} \delta(a, Z) \wedge \zeta \right) \vee 0 \\
\leq & \sup_{\mathcal{Z} \in 2^{(2^X)}} \sup_{\zeta \in \mathbb{R}^+} \sup_{\varphi \in \mathcal{Z}^A} \inf_{Z \in \mathcal{Z}} \delta\left(x, \varphi^{-1}(Z)\right) \\
= & \sup_{\mathcal{Z} \in 2^{(2^X)}} \sup_{\zeta \in \mathbb{R}^+} \sup_{\varphi \in \mathcal{Z}^A} \delta(x, A) \\
= & \, \delta(x, A).
\end{aligned}$$

On the other hand it follows from 1.8.9 that we have

$$\begin{aligned}
\delta(x, A) = & \sup_{Z \in 2^X} \sup_{\zeta \in \mathbb{R}^+} \inf_{a \in A} d_Z^\zeta(x, a) \\
\leq & \sup_{\mathcal{Z} \in 2^{(2^X)}} \sup_{\zeta \in \mathbb{R}^+} \inf_{a \in A} \sup_{Z \in \mathcal{Z}} d_Z^\zeta(x, a) \\
= & \, \delta^*(\psi(x), \psi(A)).
\end{aligned}$$

This proves that (X, δ) is indeed embedded in $\left(X^J, \delta^*\right)$.

□

3.4.2 Corollary

AP *is the epireflective and hence also the bireflective hull of pq*$\mathbf{MET}^\infty$ *in* **AP**.

□

Concept in AP	**Basic TOP-analog**	**Basic $pq\mathbf{MET}^{\infty}$-analog**
Distance (1.1.1)	Closure operator (2.1.1)	Point–set distance (3.1.1)
Adherence of $\mathcal{F}$ (1.8.34)	Adherence points of $\mathcal{F}$ (2.1.3)	$\liminf d(\mathcal{F}, \cdot)$ (3.1.3, 3.2)
Limit of $\mathcal{F}$ (1.2.1)	Limit points of $\mathcal{F}$ (2.1.3)	$\limsup d(\mathcal{F}, \cdot)$ (3.1.3, 3.2)
Approach system (1.3.1)	Neighborhoods (2.1.4)	Localized ∞pq-metrics $d(x, \cdot)$ (3.1.5)
Gauge (1.4.1)	∞pq-metrics determining coarser topologies (2.1.5)	∞pq-metric d (3.1.6)
Tower (1.5.1)	Closure operator (2.1.6)	Enlargement operators (3.1.7)
Hull operator (1.6.1)	Lower semicontinuous regularization (2.1.7)	Nonexpansive regularization (3.1.8)
Regular function frame (1.7.1)	Lower semicontinuous $[0, \infty]$-valued functions (2.1.8)	Nonexpansive $[0, \infty]$-valued functions (3.1.9)
Contraction (1.9.1, 1.9.2)	Continuous map (2.2.1)	Nonexpansive map (3.3.1)

In the table above we compare various approach concepts in **TOP** and pq**MET**$^\infty$. The purpose is not to describe in precise mathematical terms what these different approach concepts are, but rather to indicate conceptually what are the basic ideas behind these concepts. In the table, $\mathcal{F}$ stands for a filter, d stands for an ∞pq-metric, and x is a point in the underlying set.

Now that we have seen exactly in which way **TOP**, p**MET**$^\infty$, and pq**MET**$^\infty$ are embedded in **AP** we would like to make some final comments with regard to the family of bireflective bicoreflective subconstructs of **AP** which we described in 2.3.

If we consider one of the subconstructs $\mathbf{AP}_m$, for $m \neq 0$, then it follows from the definition of δ_d for an arbitrary metric space (X, d) and modulo the same abuse of notation as in 3.3.14 that the objects in $pq\mathbf{MET}^\infty \cap \mathbf{AP}_m$ are those ∞pq-metric approach spaces (X, δ_d) for which $d(X \times X) \subset \{0\} \cup [m, \infty]$. Of course this is a highly unnatural condition; it implies for instance that $\mathbb{R}$ equipped with its usual Euclidean structure is an object in none of those subconstructs.

More precisely, the only subconstruct in the family $(\mathbf{AP}_m)_{m \in [0,\infty]}$, which is a superconstruct of both **TOP** and pq**MET**$^\infty$, is $\mathbf{AP}_0 = \mathbf{AP}$. It is not even necessary to make the requirement of being a superconstruct of the whole of pq**MET**$^\infty$; p**MET**$^\infty$, **MET**, or the unique object $(\mathbb{R}, \delta_{d_E})$, as mentioned before, already suffice to force the conclusion. In this sense, **AP** is the smallest reasonable extension of **TOP** and pq**MET**$^\infty$.

4
UNIFORM APPROACH SPACES

In this chapter we will show that by embedding **TOP** and p**MET**$^\infty$ in **AP** as we have done in the foregoing chapters, we can construct arbitrary products of ∞p-metric spaces endowed with a uniquely determined distance which "distancizes" the product topology. In other words, in this chapter we solve the problem formulated under the first and the second items in the introduction. In the foregoing chapter we have seen that the epireflective hull of pq**MET**$^\infty$ is **AP** itself. In this chapter we will start by describing the epireflective hull of p**MET**$^\infty$.

4.1 THE EPIREFLECTIVE HULL OF pMET$^\infty$

The so-called epireflective hull (see Adámek *et al.* [1]) of p**MET**$^\infty$ in **AP** consists of all subspaces of products of ∞p-metric spaces. We will show that it is precisely this epireflective hull which fills in the position of the first question mark. Actually, there is a good reason why this is plausible at this point, namely the fact that **CREG** (the full subconstruct of **TOP** with objects all completely regular spaces) consists precisely of subspaces of products (in **TOP**) of ∞p-metrizable topological spaces. What we are about to show then is that this pattern repeats itself at the level of approach spaces.

Our first step is to describe products of ∞p-metric spaces in **AP**. In what follows, if $(X_j)_{j\in J}$ is a family of sets and $\prod_{j\in J} X_j$ is their product, then $\mathrm{pr}_k : \prod_{j\in J} X_j \longrightarrow X_k$ stands for the k-th projection.

4.1.1 Theorem *Let $(X_j, d_j)_{j\in J}$ be an arbitrary family of ∞p-metric spaces. The product of the family $(X_j, \delta_{d_j})_{j\in J}$ (in* **AP***) is the approach space*

$\left(\prod_{j\in J} X_j, \prod_{j\in J} \delta_{d_j}\right)$, *where* $\prod_{j\in J} \delta_{d_j}$ *is the distance generated by the gauge basis*

$$\mathcal{H} \doteqdot \left\{ \sup_{k\in K} d_k \circ (\mathrm{pr}_k \times \mathrm{pr}_k) \mid K \subset J \text{ finite} \right\}.$$

In particular, if we put $X \doteqdot \prod_{j\in J} X_j$ *and* $\delta \doteqdot \prod_{j\in J} \delta_{d_j}$ *then this distance is given by*

$$\delta(x, A) = \sup_{K\in 2^{(J)}} \inf_{a\in A} \sup_{k\in K} d_k(\mathrm{pr}_k(x), \mathrm{pr}_k(a)) \qquad \forall x \in X, \forall A \in 2^X.$$

Proof This follows from the description of initial structures in 1.10.4 and from the fact that a product in a topological construct is obtained by taking the product of the underlying sets and endowing it with the initial structure for the projections (see Appendix A or Adámek *et al.* [1]).

□

It is a fundamental aspect of the theory of approach spaces that this product, in general, is not an ∞p-metric space. More precisely, in general, there does not exist an ∞p-metric d on X such that $\delta = \delta_d$, except for finite products, as the following result shows.

4.1.2 Corollary *Let* $(X_j, d_j)_{j\in J}$ *be a finite family of* ∞p*-metric spaces. Then the product of the family* $\left(X_j, \delta_{d_j}\right)_{j\in J}$ *(in* **AP***) is an* ∞p*-metric space.*

Proof Using the same notations as in the foregoing result, it suffices to note that, for any $x \in X$ and $A \subset X$, we have

$$\begin{aligned}\delta(x, A) &= \sup_{K\in 2^{(J)}} \inf_{a\in A} \sup_{k\in K} d_k(\mathrm{pr}_k(x), \mathrm{pr}_k(a)) \\ &= \inf_{a\in A} \sup_{j\in J} d_j(\mathrm{pr}_j(x), \mathrm{pr}_j(a)) \\ &= \delta_d(x, A),\end{aligned}$$

where the ∞p-metric d is defined by

$$d \doteqdot \sup_{j\in J} d_j \circ (\mathrm{pr}_j \times \mathrm{pr}_j).$$

□

We now turn to subspaces of products of ∞p-metric spaces in **AP**. The internal characterizations which follow will be important for our further considerations. Before giving these characterizations, however, we need to introduce some more concepts and terminology.

4.1.3 Definition A gauge is called a *symmetric gauge* if it has a basis consisting of ∞p-metrics.

4.1.4 Proposition *Let (X, δ) be an approach space with symmetric gauge $\mathcal{G}$. Then $\mathcal{G} \cap p\mathrm{M}^{\infty}(X)$ is the largest basis for $\mathcal{G}$ consisting of ∞p-metrics.*

Proof If $\mathcal{H}$ is a basis for $\mathcal{G}$ consisting only of ∞p-metrics, then clearly $\mathcal{H} \subset \mathcal{G} \cap p\mathrm{M}^{\infty}(X)$ and hence $\mathcal{G} \cap p\mathrm{M}^{\infty}(X)$ too is a basis for $\mathcal{G}$. That it is the largest basis consisting of ∞p-metrics is evident.

□

4.1.5 Definition A gauge basis $\mathcal{H}$ consisting only of ∞p-metrics is called a *symmetric gauge basis.*

The foregoing result says that, if $\mathcal{G}$ is a symmetric gauge, then $\mathcal{G} \cap p\mathrm{M}^{\infty}(X)$ is its largest symmetric basis. Given a symmetric gauge basis, it is sometimes interesting to be able to work not with all the members of the gauge generated by it, but only with the ∞p-metrics in it. Thus for a symmetric gauge basis $\mathcal{H} \subset p\mathrm{M}^{\infty}(X)$ we may sometimes consider the set

$$\{d \in p\mathrm{M}^{\infty}(X) \mid \mathcal{H} \text{ dominates } d\} = \widetilde{\mathcal{H}} \cap p\mathrm{M}^{\infty}(X).$$

We call this set the *symmetric saturation* of $\mathcal{H}$ and we denote it by $\mathcal{H}^{\mathrm{s}}$. Notice that a symmetric gauge basis $\mathcal{H}$ can thus be saturated in two ways: the saturation $\widetilde{\mathcal{H}}$ taken in $pq\mathrm{M}^{\infty}(X)$ and the symmetric saturation $\mathcal{H}^{\mathrm{s}} = \widetilde{\mathcal{H}} \cap p\mathrm{M}^{\infty}(X)$ taken in $p\mathrm{M}^{\infty}(X)$. If $\mathcal{G}$ is a gauge then $\mathcal{G}^{\mathrm{s}}$ simply consists of all ∞p-metrics in $\mathcal{G}$. Hence if δ is the distance associated with $\mathcal{G}$ then it follows from 1.8.3 that $\mathcal{G}^{\mathrm{s}} = \{d \in p\mathrm{M}^{\infty}(X) \mid \delta_d \leq \delta\}$.

4.1.6 Theorem *Let (X, δ) be an approach space with gauge $\mathcal{G}$. Then the following are equivalent:*

1. *(X, δ) is a subspace of a product of ∞p-metric spaces in* **AP**,
2. *there exists a symmetric gauge basis for $\mathcal{G}$,*

3. *$\mathcal{G}^s$ is a basis for $\mathcal{G}$,*
4. *$\mathcal{G}$ is a symmetric gauge.*

Proof To prove that 1 $\Rightarrow$ 2 let $(X_j, d_j)_{j\in J}$ be an arbitrary family of ∞p-metric spaces, and suppose that (X, δ) is a subspace of $\left(\prod_{j\in J} X_j, \prod_{j\in J} \delta_{d_j}\right)$. Then it follows from 4.1.1 that

$$\mathcal{H} \doteqdot \left\{\sup_{k\in K} d_k \circ (\mathrm{pr}_k \times \mathrm{pr}_k)_{|X\times X} \mid K \subset J \text{ finite}\right\}$$

is a symmetric gauge basis for $\mathcal{G}$.

That 2 $\Leftrightarrow$ 3 follows at once from 4.1.4.

That 3 $\Rightarrow$ 4 is by definition.

To prove that 4 $\Rightarrow$ 1, let $\mathcal{H}$ be a symmetric gauge basis for $\mathcal{G}$ and consider the diagonal injection

$$\psi : (X, \delta) \longrightarrow \left(X^{\mathcal{H}}, \prod_{d\in\mathcal{H}} \delta_d\right) : x \longrightarrow (x_d \doteqdot x)_{d\in\mathcal{H}}.$$

Let $\delta^* \doteqdot \prod_{d\in\mathcal{H}} \delta_d$. For any $x \in X$ and $A \subset X$, making use of the fact that $\mathcal{H}$ is a basis for $\mathcal{G}$, we then immediately have

$$\begin{aligned}\delta^*(\psi(x), \psi(A)) &= \sup_{\mathcal{D}_0\in 2^{(\mathcal{H})}} \inf_{a\in A} \sup_{d\in\mathcal{D}_0} d(x, a)\\ &= \sup_{d\in\mathcal{H}} \inf_{a\in A} d(x, a)\\ &= \delta(x, A),\end{aligned}$$

which proves our claim.

□

Sometimes it is interesting to be able to work with bounded p-metrics. The following result shows that this is no problem.

4.1.7 Theorem *Let (X, δ) be an approach space with gauge $\mathcal{G}$. Then the following are equivalent:*

1. *(X, δ) is a subspace of a product of ∞p-metric spaces in* **AP**,
2. *there exists a symmetric gauge basis for $\mathcal{G}$ consisting of bounded p-metrics,*
3. *the set of all bounded p-metrics in $\mathcal{G}^s$ is a basis for $\mathcal{G}$.*

Proof This follows from 4.1.6. In the first place, if $\mathcal{H}$ is a symmetric gauge basis for $\mathcal{G}$ then

$$\mathcal{H}' \doteqdot \left\{ \sup_{d \in \mathcal{H}_0} (d \wedge \alpha(d)) \mid \mathcal{H}_0 \in 2^{(\mathcal{H})},\ \alpha \in]0, \infty[^{\mathcal{H}_0} \right\}$$

is a set of bounded p-metrics which is also a basis for $\mathcal{G}$. In the second place, if such a basis exists, then certainly the collection of all bounded p-metrics in $\mathcal{G}$ is a basis for $\mathcal{G}$.

□

4.1.8 Definition A subspace of a product of ∞p-metric approach spaces in **AP** will be called a *uniform approach space*. The full subconstruct consisting of all these spaces, i.e. the epireflective hull of $p\mathbf{MET}^{\infty}$ in **AP**, will be denoted by **UAP**. Structures on uniform approach spaces will also be called uniform, i.e. we will speak of a *uniform distance*, a *uniform approach system*, and so on. As before (see 1.8.32) if a uniform approach space is determined by a particular symmetric gauge basis then we will usually refer to the structures of that approach space as being *generated* by this base.

Referring to the problem explained under the first item in the introduction, the foregoing results show that **UAP** is a candidate to fill in the position of the first question mark in the diagram. It follows from the body of results of this chapter that it is the right candidate. Referring to the problem explained under the second item in the introduction, 4.1.1 in particular shows that products of metric spaces are indeed still equipped with a notion of "distance", no longer between pairs of points but rather between points and sets.

The result of 4.1.6, combined with the general descriptions obtained in the first chapter, allows us to give useful characterizations of all the other defining structures of uniform approach spaces, in the sense that we may always restrict ourselves to ∞p-metrics, should this be necessary or useful.

4.1.9 Theorem *Suppose $(X, \delta) \in |\mathbf{UAP}|$ and let δ be generated by the symmetric gauge basis $\mathcal{H}$. Then the following properties hold:*

1. the distance δ is given by

$$\delta(x, A) = \sup_{d \in \mathcal{H}} \inf_{a \in A} d(x, a) \qquad \forall x \in X, \forall A \subset X,$$

2. the family $(\mathcal{B}(x))_{x \in X}$, where

$$\mathcal{B}(x) \doteqdot \{d(x, \cdot) \mid d \in \mathcal{H}\} \qquad \forall x \in X,$$

is a basis for the approach system associated with δ,

3. *the adherence and limit operators associated with* δ *are determined by*

$$\alpha\mathcal{F}(x) = \sup_{d\in\mathcal{H}} \sup_{F\in\mathcal{F}} \inf_{y\in F} d(x,y) = \sup_{d\in\mathcal{H}} \alpha_d\mathcal{F}(x) \qquad \forall x \in X,\ \forall \mathcal{F} \in \mathbf{F}(X),$$

$$\lambda\mathcal{F}(x) = \sup_{d\in\mathcal{H}} \inf_{F\in\mathcal{F}} \sup_{y\in F} d(x,y) = \sup_{d\in\mathcal{H}} \lambda_d\mathcal{F}(x) \qquad \forall x \in X,\ \forall \mathcal{F} \in \mathbf{F}(X),$$

4. *the hull operator associated with* δ *is determined by*

$$h(\mu) = \sup_{d\in\mathcal{H}} h_d(\mu) \qquad \forall \mu \in [0,\infty]^X.$$

Proof This follows immediately from 4.1.6 and the relevant formulas proved in 1.8 and in the third chapter.

□

In particular of course the foregoing formulas hold for $\mathcal{H} = \mathcal{G}^s$.

We end this section with an internal characterization of uniform approach spaces, not unlike the characterization of completely regular topological spaces. This is no coincidence as we will see in 4.3.3 and 4.3.4. We denote by $\mathcal{K}(X)$ (respectively $\mathcal{K}^*(X)$) the set of all contractions (respectively bounded contractions) from (X,δ) to $(\mathbb{R},\delta_{d_E})$ where d_E stands for the Euclidean metric on $\mathbb{R}$.

4.1.10 Theorem *An approach space* (X,δ) *is uniform if and only if the following condition holds:* $\forall x \in X$, $\forall A \in 2^X$, $\forall \epsilon > 0$, $\forall \omega < \infty$ *there exists* $f \in \mathcal{K}(X)$ *(respectively* $f \in \mathcal{K}^*(X)$*) such that*

1. $f(x) = 0$,
2. $f_{|A} + \epsilon \geq \delta(x,A) \wedge \omega$.

Proof To show the only-if part, let $\mathcal{H}$ be a symmetric gauge basis for (X,δ) consisting of bounded pseudometrics and let $x \in X, A \in 2^X, \epsilon > 0$, and $\omega < \infty$ be fixed. Then it follows from 4.1.9 that there exists $d \in \mathcal{H}$ such that

$$\delta_d(x,A) + \epsilon \geq \delta(x,A) \wedge \omega.$$

Now, if we put $f \doteqdot d(x,\cdot)$, then it is immediately clear that $f \in \mathcal{K}^*(X)$ and that it fulfills the stated condition.

To show the if part, for each $f \in \mathcal{K}(X)$, put $d_f \doteqdot d_E \circ (f \times f)$ and consider the collection of pseudometrics $\mathcal{H} \doteqdot \{d_f \mid f \in \mathcal{K}(X)\}$. Then, from the fact that $f \in \mathcal{K}(X)$, it follows at once that d_f is a member of the gauge associated with δ. On the other hand if $x \in X, A \in 2^X, \epsilon > 0$, and $\omega < \infty$ are fixed and $f \in \mathcal{K}(X)$ is chosen according to the condition stated in the theorem, then it follows that

$$\begin{aligned}\delta_{d_f}(x, A) + \epsilon &= \inf_{a \in A} |f(x) - f(y)| + \epsilon \\ &= \inf_{a \in A} f(y) + \epsilon \\ &\geq \delta(x, A) \wedge \omega.\end{aligned}$$

Hence, for all $x \in X$ and $A \subset X$, we have that $\delta(x, A) = \sup_{d \in \mathcal{H}} \inf_{a \in A} d(x, a)$ which by 1.8.8 implies that $\mathcal{H}$ is a basis for the gauge associated with δ. Since, moreover, all members of $\mathcal{H}$ are pseudometrics, $\mathcal{H}$ is a symmetric gauge and it follows from 4.1.6 that (X, δ) is a uniform approach space.

□

Notice that in the foregoing result it suffices to impose the stated condition for closed sets A and for points x not belonging to A, which emphasizes even more the similarity with complete regularity.

We will say that a set $\mathcal{F} \subset \mathcal{K}(X)$ *generates* δ if

$$(f : (X, \delta) \longrightarrow (\mathbb{R}, \delta_{d_E}))_{f \in \mathcal{F}}$$

is an initial source.

We then obtain the following further characterizations of uniform approach spaces.

4.1.11 Proposition *For an approach space (X, δ) the following are equivalent:*

1. *(X, δ) is a uniform approach space, i.e. it is an object in* **UAP**,
2. *$\mathcal{K}^*(X)$ generates δ,*
3. *$\mathcal{K}(X)$ generates δ,*
4. *there exists a subset $\mathcal{F} \subset \mathcal{K}(X)$ which generates δ,*
5. *there exists a subset $\mathcal{F} \subset \mathcal{K}^*(X)$ which generates δ.*

Proof $1 \Rightarrow 2$. It is immediately clear that the initial distance, δ_{in}, for the source

$$(f : X \longrightarrow (\mathbb{R}, \delta_{d_E}))_{f \in \mathcal{K}^*(X)}$$

is coarser than δ, i.e. $\delta_{in} \leq \delta$. The converse inequality follows from 4.1.10, upon observing that, for any $x \in X$ and $A \subset X$,

$$\delta_{in}(x, A) = \sup_{\mathcal{H} \in 2^{(\mathcal{K}^*(X))}} \inf_{a \in A} \sup_{f \in \mathcal{H}} |f(x) - f(a)|.$$

$2 \Rightarrow 3 \Rightarrow 4$. This is trivial.

$4 \Rightarrow 5$. If $\mathcal{F} \subset \mathcal{K}(X)$ generates δ, then it is easily seen that also

$$\mathcal{F}^* \doteqdot \{(f \wedge \omega) \vee (-\omega) \mid f \in \mathcal{F},\ \omega < \infty\}$$

generates δ.

$5 \Rightarrow 1$. If $\mathcal{F} \subset \mathcal{K}^*(X)$ is a set of contractions which generates δ, then the set of p-metrics

$$\mathcal{D}_{\mathcal{F}} \doteqdot \left\{ \sup_{f \in \mathcal{H}} d_f \mid \mathcal{H} \in 2^{(\mathcal{F})} \right\},$$

where for any $f \in \mathcal{K}^*(X)$, d_f is defined by

$$d_f : X \times X \longrightarrow \mathbb{R} : (x, y) \longrightarrow |f(x) - f(y)|,$$

is a symmetric gauge which generates δ and hence, by 4.1.6, $(X, \delta) \in |\mathbf{UAP}|$.

□

4.2 CONVERGENCE IN UAP

An interesting relationship between distance and limit operator in uniform approach spaces, and which follows from the foregoing results, is the following.

4.2.1 Proposition *Suppose $(X, \delta) \in |\mathbf{UAP}|$. If $\mathcal{F}$ is a filter on X, then, for all $x, y \in X$, we have*

$$\delta(x, \{y\}) \leq \lambda\mathcal{F}(x) + \lambda\mathcal{F}(y).$$

Proof Let $x, y \in X$; then we have

$$\begin{aligned} \lambda\mathcal{F}(x) + \lambda\mathcal{F}(y) &= \sup_{d \in \mathcal{G}^\bullet} \inf_{F \in \mathcal{F}} \sup_{z \in F} d(x, z) + \sup_{d \in \mathcal{G}^\bullet} \inf_{F \in \mathcal{F}} \sup_{z \in F} d(y, z) \\ &\geq \sup_{d \in \mathcal{G}^\bullet} \inf_{F \in \mathcal{F}} \sup_{z \in F} (d(x, z) + d(y, z)) \\ &\geq \sup_{d \in \mathcal{G}^\bullet} \inf_{F \in \mathcal{F}} \sup_{z \in F} d(x, y) \\ &= \delta(x, \{y\}). \end{aligned}$$

□

As the following example shows, even in metric spaces and for sequences with a "real" limit point, the inequality in the above result is both best possible and strict in general. We mention that this example also shows that the inequality of 1.8.40 is both best possible and strict in general.

4.2.2 Example Consider the real line $\mathbb{R}$ with the usual Euclidean metric d_E, and consider the constant sequence $(x_n)_n$, where

$$x_n \doteqdot 0 \qquad \forall n \in \mathbb{N}.$$

Then it is immediately clear that, for any $x \in \mathbb{R}$,

$$\lambda_{d_E}(\langle(x_n)_n\rangle)(-x) = \lambda_{d_E}(\langle(x_n)_n\rangle)(x) = |x|$$

and

$$\delta_{d_E}(x, \{-x\}) = 2|x|,$$

and hence, with $y \doteqdot x$, the inequality reduces to $0 \leq 2|x|$. Making x arbitrarily large shows, moreover, that the bounds can be made to be arbitrarily "bad". On the other hand, taking $y \doteqdot -x$ reduces the inequality to $2|x| = 2|x|$.

Combining the foregoing result with 1.8.40 makes it possible to generalize 3.2.1 to uniform approach spaces.

4.2.3 Corollary *Suppose* $(X, \delta) \in |\mathbf{UAP}|$. *If* $\mathcal{F} \in \mathbf{F}(X)$ *and* $x \in X$ *then the following are equivalent:*

1. $\mathcal{F} \to x$ *in* $(X, \mathcal{T}_\delta)$,
2. $\lambda\mathcal{F} = \delta(\cdot, \{x\})$.

Proof That the second property implies the first one is evident and that the first property implies the second one follows at once from 1.8.40 and 4.2.1.

□

As was the case for ∞p-metric spaces, here too we can define a notion of "measure of Cauchyness" of a filter. Let $(X, \delta) \in |\mathbf{UAP}|$, let $\mathcal{H}$ be a symmetric gauge basis, and let $\mathcal{F} \in \mathbf{F}(X)$. Then we define the *$\mathcal{H}$-width* of $\mathcal{F}$ as

$$\omega_{\mathcal{H}}\mathcal{F} \doteqdot \sup_{d \in \mathcal{H}} \inf_{F \in \mathcal{F}} \operatorname{diam}_d(F).$$

This number is dependent on the particular choice of gauge basis. It is shown in example 3 of 4.4.4 that the families $\mathcal{D} \doteqdot \{d_a \mid a > 0\}$ and $\mathcal{D}' \doteqdot \{d_E\}$ (where d_E stands for the Euclidean metric on $\mathbb{R}$ and where, for each $a > 0$, d_a is defined by $d_a(x, y) \doteqdot |((-a) \vee (x \wedge a)) - ((-a) \vee (y \wedge a))|$) generate the same gauge. However, if we consider the filter $\mathcal{F}$ generated by the collection $\{[n, \infty[\mid n \in \mathbb{N}\}$, then it is immediately verified that $\omega_{\mathcal{D}}\mathcal{F} = 0$, whereas $\omega_{\mathcal{D}'}\mathcal{F} = \infty$.

One could of course always define the width with respect to $\mathcal{G}^{\mathrm{s}}$ in order to have a canonical notion dependent only on the distance. We will, however, not pursue that line of thought in the present work. The main reason for this lies in the fact that the results of 4.2.4 and 4.2.5 are actually more powerful for small symmetric gauge bases than for large ones.

Clearly $\omega_{\mathcal{D}}\mathcal{F} = 0$ if and only if $\mathcal{F}$ is a Cauchy filter in (X, d) for each $d \in \mathcal{D}$, i.e. if it is a $\mathcal{U}(\mathcal{D})$-Cauchy filter, where $\mathcal{U}(\mathcal{D})$ is the uniformity generated by $\mathcal{D}$ (see also 4.4).

The following results generalize the analogous results in 3.2.5 and 3.2.7.

4.2.4 Proposition *Suppose $(X, \delta) \in |\mathbf{UAP}|$ and let δ be generated by the symmetric gauge basis $\mathcal{H}$. If $\mathcal{F} \in \mathbf{F}(X)$, then we have*

$$\lambda\mathcal{F} \leq \alpha\mathcal{F} + \omega_{\mathcal{H}}\mathcal{F}.$$

Proof This is an immediate consequence of 3.2.5, 4.1.9, and the definition of width.

□

4.2.5 Corollary *Suppose $(X, \delta) \in |\mathbf{UAP}|$ and let δ be generated by the symmetric gauge basis $\mathcal{H}$. If $\mathcal{F} \in \mathbf{F}(X)$ is $\mathcal{U}(\mathcal{H})$-Cauchy, then $\lambda\mathcal{F} = \alpha\mathcal{F}$.*

□

We are now in a position to generalize 3.2.8 to the setting of uniform approach spaces.

4.2.6 Proposition *Suppose $(X, \delta) \in |\mathbf{UAP}|$. If $\mathcal{F} \in \mathbf{F}(X)$ is total and $x \in X$, then the following properties hold:*

1. $\alpha\mathcal{F}(x) = \inf_{y \in \mathrm{adh}_{\mathcal{T}_\delta}\mathcal{F}} d_\delta(x, y) = \delta_{d_\delta}(x, \mathrm{adh}_{\mathcal{T}_\delta}\mathcal{F})$,
2. $\lambda\mathcal{F}(x) = \sup_{y \in \mathrm{adh}_{\mathcal{T}_\delta}\mathcal{F}} d_\delta(x, y)$.

Proof Let δ be generated by the symmetric gauge basis $\mathcal{H}$.

1. If $\mathcal{F}$ is total with respect to $\mathcal{T}_\delta$, then obviously it is also total with respect to $\mathcal{T}_d$ for all $d \in \mathcal{H}$. Consequently it follows from 3.2.8 that

$$\begin{aligned}\alpha\mathcal{F}(x) &= \sup_{d\in\mathcal{H}} \alpha_d\mathcal{F}(x)\\ &= \sup_{d\in\mathcal{H}} \delta_d(x, \mathrm{adh}_{\mathcal{T}_d}\mathcal{F})\\ &\leq \sup_{d\in\mathcal{H}} \delta_d(x, \mathrm{adh}_{\mathcal{T}_\delta}\mathcal{F})\\ &\leq \delta_{d_\delta}(x, \mathrm{adh}_{\mathcal{T}_\delta}\mathcal{F}).\end{aligned}$$

To prove the other inequality, first of all notice that from 1.8.38 it follows that there exists $\mathcal{U} \in \mathrm{U}(\mathcal{F})$ such that $\alpha\mathcal{F}(x) = \lambda\mathcal{U}(x)$. Now since $\mathcal{F}$ is total, there exists $y \in \mathrm{adh}_{\mathcal{T}_\delta}\mathcal{F}$ such that $\lambda\mathcal{U}(y) = 0$, i.e. such that $\mathcal{U}$ converges to y in $\mathcal{T}_\delta$. Then it follows from 4.2.3 that $\lambda\mathcal{U}(x) = \delta(x, \{y\}) = \delta_{d_\delta}(x, \{y\})$, from which it further follows that

$$\begin{aligned}\alpha\mathcal{F}(x) &= \delta_{d_\delta}(x, \{y\})\\ &\geq \delta_{d_\delta}(x, \mathrm{adh}_{\mathcal{T}_\delta}\mathcal{F}).\end{aligned}$$

2. Making use of the fact that each ultrafilter finer than $\mathcal{F}$ converges to some point of $\mathrm{adh}_{\mathcal{T}_\delta}\mathcal{F}$, it follows from the first property, 1.8.36, and 1.8.37 that

$$\begin{aligned}\lambda\mathcal{F}(x) &= \sup_{\mathcal{U}\in\mathrm{U}(\mathcal{F})} \lambda\mathcal{U}(x)\\ &= \sup_{\mathcal{U}\in\mathrm{U}(\mathcal{F})} \delta_{d_\delta}(x, \mathrm{adh}_{\mathcal{T}_\delta}\mathcal{U})\\ &\leq \sup_{y\in\mathrm{adh}_{\mathcal{T}_\delta}\mathcal{F}} d_\delta(x, y).\end{aligned}$$

Conversely, it follows from 3.2.8 that

$$\begin{aligned}\lambda\mathcal{F}(x) &= \sup_{d\in\mathcal{H}} \lambda_d\mathcal{F}(x)\\ &= \sup_{d\in\mathcal{H}} \sup_{y\in\mathrm{adh}_{\mathcal{T}_d}\mathcal{F}} d(x, y)\\ &\geq \sup_{d\in\mathcal{H}} \sup_{y\in\mathrm{adh}_{\mathcal{T}_\delta}\mathcal{F}} d(x, y)\\ &= \sup_{y\in\mathrm{adh}_{\mathcal{T}_\delta}\mathcal{F}} d_\delta(x, y).\end{aligned}$$

□

4.2.7 Corollary *Suppose* $(X, \delta) \in |\mathbf{UAP}|$. *If* $A \subset X$ *is relatively compact and* $x \in X$ *then* $\delta(x, A) = \delta_{d_\delta}(x, \mathrm{cl}_{\mathcal{T}_\delta}(A))$.

Proof If A is relatively compact then the filter generated by $\{A\}$ is total and the result follows immediately from 4.2.6.

□

The foregoing result says that, in a uniform approach space, for a subset A of X which is relatively compact with respect to the **TOP**-bicoreflection, the distance functional δ_A is completely determined by the $p\mathbf{MET}^\infty$-bicoreflection and the **TOP**-bicoreflection.

4.3 CATEGORICAL ASPECTS

We will now see how **UAP** is embedded in **AP** and how it is related to other subconstructs of **AP**. Note that we prove slightly more than necessary, since we already know that **UAP** is epireflective in **AP**.

4.3.1 Theorem **UAP** *is a concretely bireflective subconstruct of* **AP**. *If* $(X, \delta) \in |\mathbf{AP}|$, *then its* **UAP**-*bireflection is given by*

$$id_X : (X, \delta) \longrightarrow (X, \delta^u),$$

where δ^u *is the distance generated by the gauge basis* $\mathcal{G}^s$.

Proof That $id_X : (X, \delta) \longrightarrow (X, \delta^u)$ is a contraction, by 1.9.2, follows from the fact that $\mathcal{G}^s \subset \mathcal{G}$. Now suppose that $(X', \delta') \in |\mathbf{UAP}|$, where δ' is generated by the symmetric gauge $\mathcal{G}'$, and let

$$f : (X, \delta) \longrightarrow (X', \delta')$$

be a contraction. Then, for any $d' \in (\mathcal{G}')^s$, we have $d' \circ (f \times f) \in \mathcal{G}^s$. Again, by 1.9.2 and the remark following it, this implies that

$$f : (X, \delta^u) \longrightarrow (X', \delta')$$

is a contraction.

□

4.3.2 Corollary **UAP** *is closed under the formation of limits and initial structures in* **AP**. *In particular, a product in* **AP** *of a family of uniform approach spaces is a uniform approach space and, likewise, a subspace in* **AP** *of a uniform approach space is a uniform approach space.*

Proof This follows from 4.3.1 and Adámek *et al.* [1].

□

4.3.3 Theorem *Suppose* $(X, \delta) \in |\mathbf{UAP}|$ *and let* δ *be generated by the symmetric gauge basis* $\mathcal{H}$. *Then the following hold:*

1. the **TOP**-*bicoreflection of* (X, δ) *is determined by the topology* $\mathcal{T}_\delta$ *which has as a basis for the neighborhoods the collections*

$$\{B_d(x, \epsilon) \mid d \in \mathcal{H}, \epsilon > 0\} \qquad \forall x \in X$$

and as such $\mathcal{T}_\delta$ *is a completely regular topology, uniformizable by the uniformity generated by the collection* $\mathcal{H}$ *of* ∞p-*metrics,*

2. the $\mathrm{p}\mathbf{MET}^\infty$-*bicoreflection of* (X, δ) *is determined by the* ∞p-*metric*

$$d_\delta : X \times X \longrightarrow [0, \infty] : (x, y) \longrightarrow \sup_{d \in \mathcal{H}} d(x, y).$$

Proof This follows from 2.2.8 and 3.3.6.

□

Note that in the foregoing result the symmetry of $\mathcal{H}$ is only needed to conclude that $\mathcal{T}_\delta$ is completely regular and that d_δ is an ∞p-metric. The expression for the neighborhood base of $\mathcal{T}_\delta$ and the formula for d_δ hold for any gauge basis as we have seen in 2.2.8 and 3.3.5 respectively.

By definition, **UAP** contains all ∞p-metric approach spaces. The following result shows that it also contains the right topological approach spaces.

4.3.4 Theorem *An approach space is at the same time uniform and topological if and only if it is of type* $(X, \delta_\mathcal{T})$, *for some completely regular topology* $\mathcal{T}$ *on* X. *In particular,* **CREG** *is embedded as a full subconstruct of* **UAP**.

Proof The only-if part follows from 4.3.3. To show the if part, let $\mathcal{T}$ be a completely regular topology on X. It is well known that $(X, \mathcal{T})$ can then be embedded in a product of p-metrizable topological spaces, say

$$\psi : (X, \mathcal{T}) \longrightarrow \left(\prod_{j\in J} X_j, \prod_{j\in J} \mathcal{T}_j\right).$$

It then follows from 2.2.4 that

$$\psi : (X, \delta_{\mathcal{T}}) \longrightarrow \left(\prod_{j\in J} X_j, \prod_{j\in J} \delta_{\mathcal{T}_j}\right)$$

is also an embedding. For each $j \in J$, let d_j be a p-metric which p-metrizes $\mathcal{T}_j$. Now, for each $j \in J$ also, consider the function

$$\begin{array}{cccc} \theta_j : & (X_j, \delta_{\mathcal{T}_j}) & \longrightarrow & \left(X_j^{\mathbb{N}}, \prod_{n\in\mathbb{N}} \delta_{nd_j}\right) \\ & x & \longrightarrow & (x)_{n\in\mathbb{N}} \end{array}$$

and put $\delta^j \doteqdot \prod_{n\in\mathbb{N}} \delta_{nd_j}$. For any $x \in X_j$ and $A \in 2^{X_j}$, we have

$$\begin{aligned} \delta^j(\theta_j(x), \theta_j(A)) &= \sup_{K\in 2^{(\mathbb{N})}} \inf_{a\in A} \sup_{n\in K} nd_j(x, a) \\ &= \sup_{n\in\mathbb{N}} \inf_{a\in A} nd_j(x, a) \\ &= \sup_{n\in\mathbb{N}} n \inf_{a\in A} d_j(x, a) \\ &= \sup_{n\in\mathbb{N}} n\delta_{d_j}(x, A) \\ &= \begin{cases} 0 & \delta_{d_j}(x, A) = 0, \\ \infty & \delta_{d_j}(x, A) \neq 0. \end{cases} \end{aligned}$$

Consequently, for all $j \in J$, θ_j is an embedding. Since a product of embeddings is an embedding, this proves that

$$\begin{array}{cccc} \phi : & (X, \delta_{\mathcal{T}}) & \longrightarrow & \left(\prod_{j\in J} X_j^{\mathbb{N}}, \prod_{j\in J} \prod_{n\in\mathbb{N}} \delta_{nd_j}\right) \\ & x & \longrightarrow & (\theta_j \circ pr_j \circ \psi(x))_{j\in J} \end{array}$$

is an embedding.

□

4.3.5 Theorem **CREG** *is embedded as a simultaneously bireflective and bicoreflective subconstruct of* **UAP**.

Proof That **CREG** is bireflective in **UAP** follows at once from the facts that both **TOP** (see 2.2.4) and **UAP** (see 4.3.1) are bireflectively embedded in **AP**. That **CREG** is bicoreflectively embedded in **UAP** follows from the observation that the **TOP**-bicoreflection of a **UAP**-object is completely regular (see 4.3.3). □

4.4 RELATIONSHIP WITH UNIFORM SPACES

There are several different ways in which uniform structures can be characterized. See, for example, the work by Herrlich on nearness structures [48] and by Isbell on covering uniformities [55]. Uniform spaces were originally introduced by Weil [117], making use of entourages of the diagonal and of collections of p-metrics. We will reserve the term *uniformity* for such a collection of entourages. For our purposes, in view of the link with distances, the most convenient way to introduce uniformities is via collections of ∞p-metrics. We recall the definitions (see e.g. Gillman and Jerison [42]).

A set $\mathcal{G}$ of ∞p-metrics determines a *uniform structure* if it fulfills the following properties:

1. $\mathcal{G}$ is closed under the formation of finite suprema,

2. if e is an ∞p-metric and if, for any $\epsilon > 0$, there exists $d \in \mathcal{G}$ and $\delta > 0$ such that

$$\forall x, y \in X: \; d(x,y) \le \delta \;\Rightarrow\; e(x,y) \le \epsilon,$$

then $e \in \mathcal{G}$.

The set $\mathcal{G}$ itself is often called a *uniform gauge* (see e.g. Dugundji [31]). Given a uniform gauge, the uniformity $\mathcal{U}(\mathcal{G})$ derived from $\mathcal{G}$ is given by

$$\mathcal{U}(\mathcal{G}) \doteqdot \left\{U \in 2^{X\times X} \mid \exists d \in \mathcal{G}, \exists \epsilon > 0 : \{d < \epsilon\} \subset U\right\}.$$

Any collection of ∞p-metrics determines a uniform structure, even if it fulfills no conditions at all. It suffices to note that any intersection of uniform gauges is again a uniform gauge and therefore, given a set $\mathcal{D}$ of ∞p-metrics, there always is a smallest uniform gauge $\mathcal{G}_u(\mathcal{D})$ containing $\mathcal{D}$. Thus any collection of ∞p-metrics $\mathcal{D}$ also determines a uniformity $\mathcal{U}(\mathcal{G}_u(\mathcal{D}))$, which we denote by $\mathcal{U}(\mathcal{D})$ for short.

Conversely, given a uniformity $\mathcal{U}$, the uniform gauge derived from $\mathcal{U}$ is given by

$$\mathcal{G}_u(\mathcal{U}) \doteq \left\{ d \in [0,\infty]^{X\times X} \mid d \text{ a uniformly continuous } \infty p\text{-metric} \right\},$$

where uniform continuity is understood with respect to the product uniformity $\mathcal{U} \times \mathcal{U}$ on $X \times X$ and the usual uniformity on $[0,\infty]$.

The topology underlying a uniformity $\mathcal{U}$ is denoted by $\mathcal{T}(\mathcal{U})$. This assignment generates a natural forgetful functor from **UNIF** to **CREG**. This functor, moreover, has an adjoint, called the *fine functor*, which associates with each completely regular topological space the uniform space equipped with the finest uniformity compatible with the given topology. If $\mathcal{T}$ is a completely regular topology on X, then we denote this *fine uniformity* by $\mathcal{U}_{\text{fine}}(\mathcal{T})$.

Given a uniform approach space (X,δ), we have seen in 4.1.4 that there exists a largest collection of ∞p-metrics generating δ, namely $\mathcal{G}^s$. This collection in turn determines a uniformity $\mathcal{U}(\mathcal{G}^s)$, which we will denote by $\mathcal{U}(\delta)$ for short.

We now introduce a number of functors which will help us clarify the relationship between the main constructs involved.

The first functor associates with a given uniform space the topological approach space determined by the uniform topology. In fact, via the embedding of **TOP** in **AP**, this is nothing more than the forgetful functor which associates the underlying topological space to a given uniform space.

$$\begin{array}{ccc} \mathbf{UNIF} & \overset{D}{\to} & \mathbf{UAP} \\ (X,\mathcal{U}) & \mapsto & (X,\delta_{\mathcal{T}(\mathcal{U})}) \\ f & \mapsto & f \end{array}$$

The second functor associates with a given uniform approach space the uniform space equipped with the fine uniformity generated by the underlying completely regular topology. Restricted to **CREG**, this is the fine functor.

$$\begin{array}{ccc} \mathbf{UAP} & \overset{U_1}{\to} & \mathbf{UNIF} \\ (X,\delta) & \mapsto & (X,\mathcal{U}_{\text{fine}}(\mathcal{T}_\delta)) \\ f & \mapsto & f \end{array}$$

The third functor finally associates with a given uniform approach space the uniform space equipped with the uniformity generated by the largest symmetric gauge basis for the given approach space.

$$\begin{array}{ccc} \mathbf{UAP} & \overset{U_2}{\to} & \mathbf{UNIF} \\ (X,\delta) & \rightarrowtail & (X,\mathcal{U}(\delta)) \\ f & \rightarrowtail & f \end{array}$$

Furthermore, we will denote the usual forgetful functor from **UNIF** to **TOP** (or **CREG**) by T and the bicoreflector from **AP** to **TOP** by C.

4.4.1 Theorem *1. All assignments* $\mathbf{UNIF} \overset{D}{\to} \mathbf{UAP}$, $\mathbf{UAP} \overset{U_1}{\to} \mathbf{UNIF}$ *and* $\mathbf{UAP} \overset{U_2}{\to} \mathbf{UNIF}$ *are faithful functors.*

2. The diagrams of faithful functors

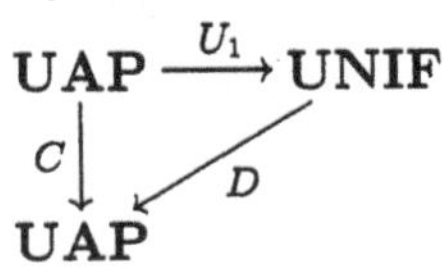

and

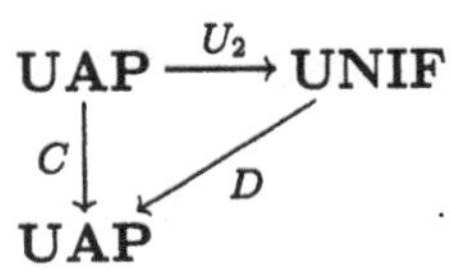

are commutative.

Proof 1. Since D is actually the forgetful functor which associates the underlying topological space to a given uniform space, D is faithful. If U_1 and U_2 are functors, then they are obviously faithful. Let (X,δ), $(X',\delta') \in |\mathbf{UAP}|$ and let $f : (X,\delta) \longrightarrow (X',\delta')$ be a contraction. To show that U_1 is a functor, it suffices to note that, by 2.2.6, $f : (X,\mathcal{T}_\delta) \longrightarrow (X',\mathcal{T}_{\delta'})$ is continuous and consequently $f : (X,\mathcal{U}_{\text{fine}}(\mathcal{T}_\delta)) \longrightarrow (X',\mathcal{U}_{\text{fine}}(\mathcal{T}_{\delta'}))$ is uniformly continuous. To show that U_2 is a functor, let $d' \in \mathcal{G}(\delta')$; then, for all $x \in X$ and $A \in 2^X$, we have

$$\begin{aligned} \delta_{d'}(f(x), f(A)) &\leq \delta'(f(x), f(A)) \\ &\leq \delta(x, A), \end{aligned}$$

which proves that $d' \circ (f \times f) \in \mathcal{G}(\delta)$. By the arbitrariness of $d' \in \mathcal{G}(\delta')$ this shows that $f : (X,\mathcal{U}(\delta)) \longrightarrow (X',\mathcal{U}(\delta'))$ is uniformly continuous.

2. This follows at once from the definitions.

□

4.4.2 Proposition *1. If (X, δ_d) is an ∞p-metric approach space then $\mathcal{U}(\delta_d)$ coincides with the uniformity generated by the ∞p-metric d.*

2. If $(X, \delta_\mathcal{T})$ is a topological uniform approach space then $\mathcal{U}(\delta_\mathcal{T})$ coincides with the fine uniformity determined by $\mathcal{T}$.

Proof Both properties follow from the corresponding results in 4.3.5.

□

4.4.3 Proposition *Given the uniform distances δ and δ' on X, the following hold:*

$$\delta = \delta' \quad \Rightarrow \quad \mathcal{U}(\delta) = \mathcal{U}(\delta') \quad \Rightarrow \quad \mathcal{T}(\delta) = \mathcal{T}(\delta').$$

Proof This follows at once from the definitions.

□

4.4.4 Examples As is the case with uniformities and topologies derived from metrics, 4.4.3 cannot be improved upon. Different uniform distances can give rise to both the same underlying topologies and the same underlying uniformities.

Also it is possible for two different collections of ∞p-metrics to generate the same distance but different uniformities.

1. Consider the real line $\mathbb{R}$ equipped with two different (not uniformly equivalent) metrics,

$$d_E : \mathbb{R} \times \mathbb{R} \longrightarrow [0, \infty] : (x, y) \longrightarrow |x - y|$$

and

$$d_0 : \mathbb{R} \times \mathbb{R} \longrightarrow [0, \infty] : (x, y) \longrightarrow |\arctan x - \arctan y|.$$

Then $\mathcal{G}(\delta_{d_E}) \neq \mathcal{G}(\delta_{d_0})$ and thus $\delta_{d_E} \neq \delta_{d_0}$ and $\mathcal{U}(\delta_{d_E}) \neq \mathcal{U}(\delta_{d_0})$. However, it follows from 4.3.3 that $\mathcal{T}_{\delta_{d_E}} = \mathcal{T}_{\delta_{d_0}}$.

2. Consider any nontopological uniform approach space (X, δ); then it follows that $(X, 2\delta)$ too is a uniform approach space and in spite of the fact that $\delta \neq 2\delta$, we have $\mathcal{U}(\delta) = \mathcal{U}(2\delta)$.

3. Consider the real line $\mathbb{R}$ and the following two collections of ∞p-metrics. First, for each real number $a > 0$ define the function

$$f_a : \mathbb{R} \longrightarrow \mathbb{R} : x \longrightarrow \begin{cases} -a & \text{if } x \leq -a, \\ x & \text{if } -a \leq x \leq a, \\ a & \text{if } a \leq x. \end{cases}$$

Second, consider the p-metrics

$$d_a(x,y) \doteqdot |f_a(x) - f_a(y)| \qquad \forall x,y \in \mathbb{R},\ \forall a > 0,$$

and let $\mathcal{D} \doteqdot \{d_a \mid a > 0\}$ and $\mathcal{D}' \doteqdot \{d_E\}$, where d_E stands for the Euclidean metric.

Then we have that both $\mathcal{D}$ and $\mathcal{D}'$ are bases for the same symmetric gauge, but $\mathcal{U}(\mathcal{D}) \neq \mathcal{U}(\mathcal{D}')$.

The results of the foregoing sections now allow us to complete our diagram of categories. In all three of the columns the transition from the first row to the second row consists in forgetting the numerical information and only retaining uniform information on a set-theoretical level. The transition from the second row to the last row then consists in forgetting the uniform information and only retaining the local information. In all three of the rows the transition from the first column to the second column consists in going from a single ∞p-metric to "arbitrary" families of ∞p-metrics and the transition from the second column to the last column consists in going from ∞p-metrics to ∞pq-metrics.

$$\begin{array}{ccccc}
p\mathbf{MET}^{\infty} & \longrightarrow & \mathbf{UAP} & \longrightarrow & \mathbf{AP} \\
\downarrow & & \downarrow & & \downarrow \\
pm\mathbf{UNIF} & \longrightarrow & \mathbf{UNIF} & \longrightarrow & q\mathbf{UNIF} \\
\downarrow & & \downarrow & & \downarrow \\
pm\mathbf{TOP} & \longrightarrow & \mathbf{CREG} & \longrightarrow & \mathbf{TOP}
\end{array}$$

Notice that in this diagram, all horizontal arrows represent the canonical functorial embeddings and all vertical arrows represent the canonical forgetful functors. The composition of the three pairs of vertical arrows actually represents (restrictions of) the **TOP**-bicoreflector in **AP**.

5
CANONICAL EXAMPLES

In the foregoing chapters we have studied three fundamental classes of approach spaces: topological spaces, metric spaces, and uniform approach spaces. The fact that **TOP** is simultaneously embedded as a bireflective and bicoreflective subconstruct means that no interesting new types of objects can be obtained from topological spaces making use of categorical constructions in **AP**. The situation for $pq\mathbf{MET}^{\infty}$ and $p\mathbf{MET}^{\infty}$, as we have seen, is quite different. The fact that the latter is not reflectively embedded in **AP** allowed us to construct its epireflective hull, the construct **UAP** of uniform approach spaces. In the foregoing chapters we were mainly concerned with filling in the positions created by the two question marks in our diagram. In the present chapter we would like to demonstrate that in various fields of mathematics, examples of natural approach spaces and traces of concepts and results which find their natural environment in the setting of approach spaces can be found.

We will consider several well-known types of uniformizable topological spaces, all of which can naturally be viewed as uniform approach spaces. Thus we will study spaces of probability measures, spaces of functions between metric spaces, hyperspaces of metric spaces, probabilistic metric spaces, and spaces of random variables.

In all of these examples we will be able to define a canonical uniform approach structure, in such a way that the topological coreflection coincides with some classical topology and that the metric coreflection coincides with some classical metric. In all cases we will give arguments, specific for the type of space, why the approach structure is natural and worthwhile studying. Sometimes part of the argumentation has to be deferred to a later stage when we have more machinery at our disposal.

5.1 SPACES OF MEASURES

Let $(X, \mathcal{T})$ be a separable metrizable topological space, and let d stand for a fixed metric generating the topology. $\mathcal{T}^*$ will stand for the collection of all closed sets in X, $\mathcal{B}$ will denote the collection of Borel sets in X, and $\mathcal{P}$ will denote the set of all probability measures on X. One of the most important and most widely used structures on $\mathcal{P}$ is the so-called *weak topology*, which we denote by $\mathcal{W}$, Billingsley [13], Parthasarathy [97]. We recall that this topology is determined by any of the following four different but equivalent bases for the neighborhoods. Let $P \in \mathcal{P}$.

1. *Base determined by the open sets in X.*

$$\mathcal{B}^0_{\mathcal{W}}(P) \doteqdot \left\{V^0(P, \mathcal{G}, \epsilon) \mid \mathcal{G} \in 2^{(\mathcal{T})}, \epsilon > 0\right\},$$

where, for any $\mathcal{G} \in 2^{(\mathcal{T})}$ and $\epsilon > 0$,

$$V^0(P, \mathcal{G}, \epsilon) \doteqdot \{Q \in \mathcal{M} \mid \forall G \in \mathcal{G} : Q(G) > P(G) - \epsilon\}.$$

2. *Base determined by the closed sets in X.*

$$\mathcal{B}^1_{\mathcal{W}}(P) \doteqdot \left\{V^1(P, \mathcal{F}, \epsilon) \mid \mathcal{F} \in 2^{(\mathcal{T}^*)}, \epsilon > 0\right\},$$

where, for any $\mathcal{F} \in 2^{(\mathcal{T}^*)}$ and $\epsilon > 0$,

$$V^1(P, \mathcal{F}, \epsilon) \doteqdot \{Q \in \mathcal{M} \mid \forall F \in \mathcal{F} : Q(F) < P(F) + \epsilon\}.$$

3. *Base determined by continuous functions on X.* Let $\mathcal{C}(X, [0, 1])$ be the set of all continuous functions from X to the unit interval; then we let

$$\mathcal{B}^2_{\mathcal{W}}(P) \doteqdot \left\{V^2(P, \mathcal{H}, \epsilon) \mid \mathcal{H} \in 2^{(\mathcal{C}(X,[0,1]))}, \epsilon > 0\right\},$$

where, for any $\mathcal{H} \in 2^{(\mathcal{C}(X,[0,1]))}$ and $\epsilon > 0$,

$$V^2(P, \mathcal{H}, \epsilon) \doteqdot \left\{Q \in \mathcal{M} \mid \forall f \in \mathcal{H} : \left|\int f \mathrm{d}P - \int f \mathrm{d}Q\right| < \epsilon\right\}.$$

4. *Base determined by uniformly continuous functions on X.* Let $\mathcal{U}(X, [0, 1])$ be the set of all uniformly continuous functions from X to the unit interval; then we let

$$\mathcal{B}^3_{\mathcal{W}}(P) \doteqdot \left\{V^3(P, \mathcal{H}, \epsilon) \mid \mathcal{H} \in 2^{(\mathcal{U}(X,[0,1]))}, \epsilon > 0\right\},$$

where, for any $\mathcal{H} \in 2^{(\mathcal{U}(X,[0,1]))}$ and $\epsilon > 0$,

$$V^3(P,\mathcal{H},\epsilon) \doteqdot \left\{Q \in \mathcal{M} \mid \forall f \in \mathcal{H} : \left|\int f\mathrm{d}P - \int f\mathrm{d}Q\right| < \epsilon\right\}.$$

A neighborhood $V^0(P,\mathcal{G},\epsilon)$ for instance contains all probability measures which on the open sets of $\mathcal{G}$ have mass almost (up to ϵ) as "big" as P. However, no distinction is made as to the precise amount that the mass "almost" exceeds that of P.

We can go about it in a more subtle way by considering the following *pq*-metrics:

1. *pq-metrics determined by the open sets in* X. For any $\mathcal{G} \in 2^{(\mathcal{T})}$, we let

$$d_0^{\mathcal{G}} : \mathcal{P} \times \mathcal{P} \longrightarrow [0,\infty] : (P,Q) \longrightarrow \sup_{G \in \mathcal{G}} (P(G) - Q(G)) \vee 0.$$

It is easily seen that this is a *pq*-metric and that the collection of *pq*-metrics

$$\mathcal{D}_0 \doteqdot \left\{d_0^{\mathcal{G}} \mid \mathcal{G} \in 2^{(\mathcal{T})}\right\}$$

is closed under the formation of finite suprema. Moreover,

$$V^0(P,\mathcal{G},\epsilon) = B_{d_0^{\mathcal{G}}}(P,\epsilon).$$

2. *pq-metrics determined by the closed sets in* X. For any $\mathcal{F} \in 2^{(\mathcal{T}^*)}$, we let

$$d_1^{\mathcal{F}} : \mathcal{P} \times \mathcal{P} \longrightarrow [0,\infty] : (P,Q) \longrightarrow \sup_{F \in \mathcal{F}} (Q(F) - P(F)) \vee 0.$$

It is also easily seen that this is a *pq*-metric and that the collection of *pq*-metrics

$$\mathcal{D}_1 \doteqdot \left\{d_1^{\mathcal{F}} \mid \mathcal{F} \in 2^{(\mathcal{T}^*)}\right\}$$

is closed under the formation of finite suprema. Moreover,

$$V^1(P,\mathcal{F},\epsilon) = B_{d_1^{\mathcal{F}}}(P,\epsilon).$$

Notice also that $d_1^{\mathcal{F}} = d_0^{\mathcal{G}}$ if $\mathcal{G} = \{X \setminus F \mid F \in \mathcal{F}\}$.

3. *p-metrics determined by continuous functions on* X. For any finite set $\mathcal{H}$ of continuous functions on X, we let

$$d_2^{\mathcal{H}} : \mathcal{P} \times \mathcal{P} \longrightarrow [0,\infty] : (P,Q) \longrightarrow \sup_{f \in \mathcal{H}} \left|\int f\mathrm{d}P - \int f\mathrm{d}Q\right|.$$

Now it is clear that this is a p-metric and that the collection of p-metrics

$$\mathcal{D}_2 \doteqdot \left\{ d_2^{\mathcal{H}} \mid \mathcal{H} \in 2^{(\mathcal{C}(X,[0,1]))} \right\}$$

too is closed under the formation of finite suprema. Moreover, we again have

$$V^2(P, \mathcal{H}, \epsilon) = B_{d_2^{\mathcal{H}}}(P, \epsilon).$$

4. *p-metrics determined by uniformly continuous functions on X.* For any finite set $\mathcal{H}$ of uniformly continuous functions on X, we let

$$d_3^{\mathcal{H}} : \mathcal{P} \times \mathcal{P} \longrightarrow [0, \infty] : (P, Q) \longrightarrow \sup_{f \in \mathcal{H}} \left| \int f \mathrm{d}P - \int f \mathrm{d}Q \right|.$$

Again it is clear that this is a p-metric and that the collection of p-metrics

$$\mathcal{D}_3 \doteqdot \left\{ d_3^{\mathcal{H}} \mid \mathcal{H} \in 2^{(\mathcal{U}(X,[0,1]))} \right\}$$

is closed under the formation of finite suprema. Moreover, we again have that

$$V^3(P, \mathcal{H}, \epsilon) = B_{d_3^{\mathcal{H}}}(P, \epsilon).$$

Being closed under the formation of finite suprema, the collections $\mathcal{D}_0$ and $\mathcal{D}_1$ are gauge bases and the collections $\mathcal{D}_2$ and $\mathcal{D}_3$ are symmetric gauge bases. Consequently, they all generate distances.

5.1.1 Proposition *The gauge bases $\mathcal{D}_0, \mathcal{D}_1$ and the symmetric gauge bases $\mathcal{D}_2, \mathcal{D}_3$ generate the same distance.*

Proof That $\mathcal{D}_0$ and $\mathcal{D}_1$ generate the same distance follows immediately using complementation. To show that $\mathcal{D}_1$ and $\mathcal{D}_2$ generate the same distance it suffices to show the following two facts.

Fact 1: $\forall P \in \mathcal{P}$, $\forall F$ closed, $\forall \epsilon > 0$ $\exists f \in \mathcal{C}(X, [0,1])$ such that

$$\forall Q \in \mathcal{P} : \quad (Q(F) - P(F)) \vee 0 \leq \left| \int f \mathrm{d}P - \int f \mathrm{d}Q \right| + \epsilon.$$

Indeed, by regularity of the probability measure P it follows that there exists $\gamma > 0$ such that

$$P\left(F^{(\gamma)}\right) \leq P(F) + \epsilon.$$

Define $f \in \mathcal{C}(X,[0,1])$ by

$$f(x) \doteqdot \left(1 - \frac{1}{\gamma}d(x,F)\right) \vee 0 \qquad \forall x \in X.$$

Then it follows that

$$\begin{aligned} Q(F) - P(F) &\leq \int f\mathrm{d}Q - P(F) \\ &\leq \int f\mathrm{d}Q - \left(P\left(F^{(\gamma)}\right) - \epsilon\right) \\ &\leq \left(\int f\mathrm{d}Q - \int f\mathrm{d}P\right) + \epsilon. \end{aligned}$$

Fact 2: $\forall f \in \mathcal{C}(X,[0,1]), \forall \epsilon > 0\ \exists \mathcal{F} \in 2^{(\mathcal{T}^*)}$ such that

$$\forall P, Q \in \mathcal{P} : \left|\int f\mathrm{d}P - \int f\mathrm{d}Q\right| \leq \sup_{F\in\mathcal{F}} (Q(F) - P(F)) \vee 0 + \epsilon.$$

Indeed, choose $k \in \mathbb{N}_0$ such that $\frac{1}{k} \leq \epsilon$ and, for all $i \in \{1,\ldots,k\}$, let

$$F_i \doteqdot \left\{\frac{i}{k} \leq f\right\}$$

and let $\mathcal{F} \doteqdot \{F_i \mid i \in \{1,\ldots,k\}\}$; then it is easily seen that, for any $P \in \mathcal{P}$,

$$\frac{1}{k}\sum_{F\in\mathcal{F}} P(F) \leq \int f\mathrm{d}P \leq \frac{1}{k} + \frac{1}{k}\sum_{F\in\mathcal{F}} P(F).$$

Consequently, for any $P, Q \in \mathcal{P}$, if we assume that $\int f\mathrm{d}Q \leq \int f\mathrm{d}P$, we have

$$\begin{aligned} \int f\mathrm{d}P - \int f\mathrm{d}Q &\leq \frac{1}{k} + \frac{1}{k}\sum_{F\in\mathcal{F}} P(F) - \frac{1}{k}\sum_{F\in\mathcal{F}} Q(F) \\ &\leq \epsilon + \frac{1}{k}\sum_{F\in\mathcal{F}} (P(F) - Q(F)) \vee 0 \\ &\leq \epsilon + \sup_{F\in\mathcal{F}} (P(F) - Q(F)) \vee 0. \end{aligned}$$

That $\mathcal{D}_3$ too generates the same distance follows from noting that the function f constructed in the first part of the foregoing proof is in fact uniformly continuous.

□

We will indicate the structures of the uniform approach space determined by any of the collections $\mathcal{D}_0, \mathcal{D}_1, \mathcal{D}_2$, and $\mathcal{D}_3$ by the use of w as an index. Thus

δ_w stands for the distance, λ_w for the limit operator, and so on. We will later see that the distance δ_w, in a natural way, "distancizes" the weak topology. For this reason we will refer to it as the *weak distance*, and similarly we will call the associated approach system the *weak approach system* and so on.

Before we describe the most interesting of these approach structures we first give some other ways of introducing δ_w.

In [114] Topsøe showed that the weak topology on $\mathcal{P}$ is the initial topology for the source

$$(\omega_G : \mathcal{P} \longrightarrow ([0,\infty], \mathcal{T}_{\mathbb{P}}) : P \longrightarrow P(G))_{G \in \mathcal{T}},$$

where $\mathcal{T}_{\mathbb{P}}$ stands for the so-called right-limit topology on $[0,\infty]$ (see 2.2.9). As we have seen in 2.2.9, this is the topology of the **TOP**-bicoreflection of the space $([0,\infty], \delta_{\mathbb{P}})$. The next result shows that Topsøe's theorem also holds in the setting of approach spaces.

5.1.2 Proposition *The weak distance δ_w on $\mathcal{P}$ is initial for the source*

$$(\omega_G : \mathcal{P} \longrightarrow ([0,\infty], \delta_{\mathbb{P}}) : P \longrightarrow P(G))_{G \in \mathcal{T}}.$$

Proof It follows from 1.8.33 that, for all $a \in [0,1]$,

$$\mathcal{A}_{\mathbb{P}}(a) = \left\{\varphi \in [0,\infty]^{[0,\infty]} \mid \varphi \leq d_{\mathbb{P}}(a,\cdot)\right\}.$$

Consequently, it follows from 1.10.1 that the approach system $(\mathcal{A}_w(P))_{P \in \mathcal{P}}$ is generated by

$$\{d_{\mathbb{P}}(P(G),\cdot) \circ \omega_G \mid G \in \mathcal{T}\} \qquad \forall P \in \mathcal{P},$$

and since $d_{\mathbb{P}}(P(G),\cdot) \circ \omega_G = d_0^{\{G\}}(P,\cdot)$, this proves our claim.

□

5.1.3 Proposition *The distance δ_w on $\mathcal{P}$ is given by the following formulas. For any $P \in \mathcal{P}$ and $\mathcal{N} \in 2^{\mathcal{P}}$:*

$$\begin{aligned}\delta_w(P,\mathcal{N}) &= \sup_{\mathcal{G} \in 2^{(\mathcal{T})}} \inf_{Q \in \mathcal{N}} d_0^{\mathcal{G}}(P,Q)\\ &= \sup_{\mathcal{F} \in 2^{(\mathcal{T}^*)}} \inf_{Q \in \mathcal{N}} d_1^{\mathcal{F}}(P,Q)\\ &= \sup_{\mathcal{H} \in 2^{(C(X,[0,1]))}} \inf_{Q \in \mathcal{N}} d_2^{\mathcal{H}}(P,Q)\\ &= \sup_{\mathcal{H} \in 2^{(\mathcal{U}(X,[0,1]))}} \inf_{Q \in \mathcal{N}} d_3^{\mathcal{H}}(P,Q).\end{aligned}$$

Proof This follows from 1.8.5, 4.1.9, and 5.1.1.

□

5.1.4 Proposition *The limit operator λ_w on $\mathcal{P}$ is given by the following formulas. For any $P \in \mathcal{P}$ and any filter Ψ in $\mathcal{P}$:*

$$\begin{aligned}\lambda_w(\Psi)(P) &= \sup_{G\in\mathcal{T}} \inf_{\mathcal{F}\in\Psi} \sup_{Q\in\mathcal{F}} (P(G) - Q(G)) \vee 0 \\ &= \sup_{F\in\mathcal{T}^*} \inf_{\mathcal{F}\in\Psi} \sup_{Q\in\mathcal{F}} (Q(F) - P(F)) \vee 0 \\ &= \sup_{f\in\mathcal{C}(X,[0,1])} \inf_{\mathcal{F}\in\Psi} \sup_{Q\in\mathcal{F}} \left| \int f dP - \int f dQ \right| \\ &= \sup_{f\in\mathcal{U}(X,[0,1])} \inf_{\mathcal{F}\in\Psi} \sup_{Q\in\mathcal{F}} \left| \int f dP - \int f dQ \right|.\end{aligned}$$

Proof From 1.8.31, 1.10.5, and 5.1.2 it immediately follows that

$$\begin{aligned}\lambda_w(\Psi)(P) &= \sup_{G\in\mathcal{T}} \lambda_{\mathbb{P}}(\text{stack}\,\omega_G(\Psi))(\omega_G(P)) \\ &= \sup_{G\in\mathcal{T}} \inf_{\mathcal{F}\in\Psi} \sup_{Q\in\mathcal{F}} (P(G) - Q(G)) \vee 0.\end{aligned}$$

The other formulas are obtained in precisely the same way.

□

The foregoing result acquires a more usual form when we work with sequences instead of with filters.

5.1.5 Corollary *For any sequence $(P_n)_n$ in $\mathcal{P}$ and any $P \in \mathcal{P}$:*

$$\begin{aligned}\lambda_w(\langle(P_n)_n\rangle)(P) &= \sup_{G\in\mathcal{T}} \limsup_{n\to\infty} (P(G) - P_n(G)) \vee 0 \\ &= \sup_{F\in\mathcal{T}^*} \limsup_{n\to\infty} (P_n(F) - P(F)) \vee 0 \\ &= \sup_{f\in\mathcal{C}(X,[0,1])} \limsup_{n\to\infty} \left| \int f dP_n - \int f dP \right| \\ &= \sup_{f\in\mathcal{U}(X,[0,1])} \limsup_{n\to\infty} \left| \int f dP_n - \int f dP \right|.\end{aligned}$$

□

5.1.6 Proposition *The tower $\mathfrak{t}_w$ on $\mathcal{P}$ is given by the following formulas. For any $\mathcal{N} \in 2^{\mathcal{P}}$ and $\epsilon \in \mathbb{R}^+$:*

$$\begin{aligned}(\mathfrak{t}_w)_\epsilon(\mathcal{N}) &= \left\{P \in \mathcal{P} \mid \forall \mathcal{G} \in 2^{(\mathcal{T})}, \forall \theta > \epsilon : \ V^0(P,\mathcal{G},\theta) \cap \mathcal{N} \neq \emptyset\right\} \\ &= \left\{P \in \mathcal{P} \mid \forall \mathcal{F} \in 2^{(\mathcal{T}^*)}, \forall \theta > \epsilon : \ V^1(P,\mathcal{F},\theta) \cap \mathcal{N} \neq \emptyset\right\} \\ &= \left\{P \in \mathcal{P} \mid \forall \mathcal{H} \in 2^{(\mathcal{C}(X,[0,1]))}, \forall \theta > \epsilon : \ V^2(P,\mathcal{H},\theta) \cap \mathcal{N} \neq \emptyset\right\} \\ &= \left\{P \in \mathcal{P} \mid \forall \mathcal{H} \in 2^{(\mathcal{U}(X,[0,1]))}, \forall \theta > \epsilon : \ V^3(P,\mathcal{H},\theta) \cap \mathcal{N} \neq \emptyset\right\}.\end{aligned}$$

Proof From 5.1.3 we obtain

$$\begin{aligned}&\delta_w(P,\mathcal{N}) \leq \epsilon \\ &\Leftrightarrow \forall \mathcal{G} \in 2^{(\mathcal{T})}, \forall \theta > \epsilon, \exists Q \in \mathcal{N} : \ d_0^{\mathcal{G}}(P,Q) < \theta \\ &\Leftrightarrow \forall \mathcal{G} \in 2^{(\mathcal{T})}, \forall \theta > \epsilon : \ V^0(P,\mathcal{G},\theta) \cap \mathcal{N} \neq \emptyset.\end{aligned}$$

The proof of the other equalities is precisely the same.

□

We now proceed with the description of the **TOP**-bicoreflection and the **pMET**$^\infty$-bicoreflection of $(\mathcal{P},\delta_w)$.

5.1.7 Theorem *The **TOP**-bicoreflection of $(\mathcal{P},\delta_w)$ is $(\mathcal{P},\mathcal{W})$ where $\mathcal{W}$ stands for the weak topology on $\mathcal{P}$.*

Proof This follows from 2.2.8 since, for any $\mathcal{G} \in 2^{(\mathcal{T})}$ and $\epsilon > 0$,

$$V^0(P,\mathcal{G},\epsilon) = B_{d_0^{\mathcal{G}}}(P,\epsilon).$$

□

As a corollary to the foregoing results we obtain the well known so-called *Portmanteau theorem*, (see e.g. Billingsley [13]).

5.1.8 Corollary (Portmanteau theorem) *For any sequence $(P_n)_n$ in $\mathcal{P}$ and any $P \in \mathcal{P}$, the following are equivalent:*

1. $(P_n)_n \to P$ *in the weak topology,*
2. $\liminf\limits_{n\to\infty} P_n(G) \geq P(G)$ $\qquad \forall G$ *open,*
3. $\limsup\limits_{n\to\infty} P_n(F) \leq P(F)$ $\qquad \forall F$ *closed,*
4. $\lim\limits_{n\to\infty} \int f \mathrm{d}P_n = \int f \mathrm{d}P$ $\qquad \forall f \in \mathcal{C}(X, [0,1])$,
5. $\lim\limits_{n\to\infty} \int f \mathrm{d}P_n = \int f \mathrm{d}P$ $\qquad \forall f \in \mathcal{U}(X, [0,1])$.

Proof This follows at once from 2.2.8, 5.1.5, and 5.1.7.

□

We recall that the $\mathbf{L}^1$-metric (also called the *total variation metric*) on $\mathcal{P}$ is given by $d(P,Q) \doteqdot \sup\limits_{B\in\mathcal{B}} |P(B) - Q(B)|$, for all $P, Q \in \mathcal{P}$ (see e.g. Parthasarathy and Steerneman [98]).

5.1.9 Theorem *The* pMET$^\infty$*-bicoreflection of* $(\mathcal{P}, \delta_w)$ *is* $(\mathcal{P}, d)$*, where d is the* $\mathbf{L}^1$*-metric on* $\mathcal{P}$.

Proof From 3.3.6 we obtain that

$$\begin{aligned}\delta_w(P, \{Q\}) &= \sup_{\mathcal{G}\in 2^{(\mathcal{T})}} \sup_{G\in\mathcal{G}} d_0^{\mathcal{G}}(P,Q) \\ &= \sup_{G\in\mathcal{T}} d_0^{\{G\}}(P,Q).\end{aligned}$$

The result now follows from the fact that probability measures on a metric space are regular, see e.g. Parthasarathy [97], and hence $\sup\limits_{G\in\mathcal{T}} d_0^{\{G\}}(P,Q) = \sup\limits_{B\in\mathcal{B}} |P(B) - Q(B)|$.

□

Another fundamental fact about the weak topology is that if $f : X \longrightarrow Y$ is a continuous function then its canonical extension $\hat{f} : \mathcal{P}(X) \longrightarrow \mathcal{P}(Y)$ defined by $\hat{f}(P)(B) \doteqdot P(f^{-1}(B))$, for all $B \in \mathcal{B}(Y)$, is continuous with respect to the weak topologies. Here too we are able to prove a stronger analog.

5.1.10 Proposition *If $f : X \longrightarrow Y$ is a continuous function and we equip $\mathcal{P}(X)$ and $\mathcal{P}(Y)$ with the weak distances, then $\hat{f} : \mathcal{P}(X) \longrightarrow \mathcal{P}(Y)$ is a contraction.*

Proof Let $\mathcal{G}$ be a finite collection of open sets in Y, and let $P, Q \in \mathcal{P}(X)$; then

$$\begin{aligned} d_0^{\mathcal{G}}\left(\hat{f}(P), \hat{f}(Q)\right) &= \sup_{G \in \mathcal{G}} \left(P(f^{-1}(G)) - Q(f^{-1}(G))\right) \vee 0 \\ &= d_0^{\mathcal{H}}(P, Q), \end{aligned}$$

where $\mathcal{H} \doteqdot \{f^{-1}(G) \mid G \in \mathcal{G}\}$. This proves that

$$d_0^{\mathcal{G}}\left(\hat{f}(P), \cdot\right) \circ \hat{f} = d_0^{\mathcal{H}}(P, \cdot) \in \mathcal{A}_w(P),$$

which by 1.9.2, and the remark following it, proves our claim.

□

5.1.11 Corollary *If $f : X \longrightarrow Y$ is continuous then $\hat{f} : \mathcal{P}(X) \longrightarrow \mathcal{P}(Y)$ is continuous with respect to the weak topologies and nonexpansive with respect to the $\mathbf{L}^1$-metrics.*

□

Finally, to conclude this example, we would like to calculate the distance between an arbitrary probability measure and a specific set of probability measures. It is well known that for the weak topology, X is embedded as a closed subspace of $\mathcal{P}$ by

$$\Delta : X \longrightarrow \mathcal{P} : x \longrightarrow P_x,$$

where P_x is the Dirac measure in x (i.e. $P_x(B) = 1$ if $x \in B$ and $P_x(B) = 0$ if $x \notin B$ for all $B \in \mathcal{B}$). In the setting of approach spaces, being closed means that if the distance from a probability measure P to the set $\Delta(X)$ is zero, then P itself is a Dirac measure. Let us see what this becomes for the weak distance.

5.1.12 Proposition *For any $P \in \mathcal{P}$, we have*

$$\delta_w(P, \Delta(X)) = 1 - \sup_{x \in X} P(\{x\}).$$

Proof We restrict ourselves to the case where P has a countably infinite set of atoms. The other cases are similar and more simple and are left to the reader. Let

α be the P-measure of an atom $a \in X$ of largest measure. Then we have

$$\begin{aligned}\delta_w(P, \Delta(X)) &\leq \delta_w(P, \{P_a\}) \\ &= \sup_{\mathcal{G} \in 2^{(\mathcal{T})}} \sup_{G \in \mathcal{G}} (P(G) - P_a(G)) \vee 0 \\ &= \sup_{\mathcal{G} \in \mathcal{T}} (P(G) - P_a(G)) \vee 0 \\ &= P(X \setminus \{a\}) - P_a(X \setminus \{a\}) \\ &= 1 - \alpha.\end{aligned}$$

Conversely, if $\mathcal{G} \in 2^{(\mathcal{T})}$ then, in the case where $\cap\mathcal{G} \neq \emptyset$, we have

$$\inf_{x \in X} \sup_{G \in \mathcal{G}} (P(G) - P_x(G)) \vee 0 = 0.$$

Consequently

$$\begin{aligned}\delta_w(P, \Delta(X)) &= \sup_{\mathcal{G} \in 2^{(\mathcal{T})}} \inf_{x \in X} \sup_{G \in \mathcal{G}, x \notin G} P(G) \\ &= \sup_{\mathcal{G} \in 2^{(\mathcal{T})}} \inf_{H \in \mathcal{G}} \inf_{x \in X \setminus H} \sup_{G \in \mathcal{G}, x \notin G} P(G) \\ &\geq \sup_{\mathcal{G} \in 2^{(\mathcal{T})}, \cap\mathcal{G} = \emptyset} \inf_{H \in \mathcal{G}} \inf_{x \in X \setminus H} P(H) \\ &\geq \sup_{\mathcal{G} \in 2^{(\mathcal{T})}, \cap\mathcal{G} = \emptyset} \inf_{G \in \mathcal{G}} P(G).\end{aligned}$$

Now let us suppose that the set of P-atoms $\{a_n \mid n \in \mathbb{N}\}$ is ordered in the sense that $P(\{a_{n+1}\}) \leq P(\{a_n\})$, for all $n \in \mathbb{N}$. Then $\alpha = P(\{a_0\})$. Let $\beta \doteqdot \max\{\{P(\{x\}) \mid x \in X\} \setminus \{\alpha\}\}$, i.e. β is the P-measure of a second largest P-atom. Let $\epsilon \in]0, \alpha - \beta[$ and take $n_0 \in \mathbb{N}$ such that

$$\sum_{n > n_0} P(\{a_n\}) < \frac{\epsilon}{2}.$$

Next choose $r > 0$ such that, for all $n \in \{0, \dots, n_0\}$,

$$P(B^*(a_n, r)) \leq P(\{a_n\}) + \epsilon.$$

Let $Y \doteqdot X \setminus \{a_n \mid n \in \mathbb{N}\}$ and partition Y into disjoint Borel sets $D_0, \dots, D_{n_1}$ such that, for all $i \in \{0, \dots, n_1\}$,

$$P(D_i) < \beta.$$

Next choose $\theta > 0$ such that $n_1\theta \leq \frac{\epsilon}{2}$. Since P is regular we can find closed sets $K_i \subset D_i$, for all $i \in \{0, \ldots, n_1\}$, such that

$$P(K_i) \leq P(D_i) \leq P(K_i) + \theta.$$

Since the sets $K_0, \ldots, K_{n_1}, \{a_n \mid n \leq n_0\}$, are pairwise disjoint and closed it follows from the normality of X and the regularity of P that we can find open sets $O_i \supset K_i$ such that the following properties are fulfilled:

(i) $O_0, \ldots, O_{n_1}$ are pairwise disjoint,

(ii) $\forall i \in \{0, \ldots, n_1\}$: $P(K_i) \leq P(O_i) \leq P(K_i) + \theta$,

(iii) $\forall i \in \{0, \ldots, n_1\}$: $O_i \cap \{a_n \mid n \leq n_0\} \neq \emptyset$.

Now consider the following finite collections of open sets in X:

$$G_i \doteqdot \bigcup_{j=0, j\neq i}^{n_1} O_j \cup \bigcup_{k=0}^{n_0} B\left(a_k, \frac{r}{2}\right) \qquad i \in \{0, \ldots, n_1\},$$

$$H_i \doteqdot X \setminus B^*(a_i, r) \qquad i \in \{0, \ldots, n_0\}.$$

By construction we have

$$\left(\bigcap_{i=0}^{n_1} G_i\right) \cap \left(\bigcap_{i=0}^{n_0} H_i\right) = \emptyset.$$

For each $i \in \{0, \ldots, n_1\}$, we further have

$$\begin{aligned}
P(G_i) &\geq P\left(\left(\bigcup_{j=0, j\neq i}^{n_1} O_j\right) \cup \{a_n \mid n \leq n_0\}\right) \\
&= P\left(\bigcup_{j=0, j\neq i}^{n_1} O_j\right) + \sum_{n \leq n_0} P(\{a_n\}) \\
&= P\left(\bigcup_{j=0, j\neq i}^{n_1} O_j\right) + 1 - P(Y) - \sum_{n_0 < n} P(\{a_n\}) \\
&\geq P\left(\bigcup_{j=0, j\neq i}^{n_1} D_j\right) - \frac{\epsilon}{2} + 1 - P(Y) - \frac{\epsilon}{2} \\
&= 1 - \epsilon - P(D_i) \\
&\geq 1 - \epsilon - \beta \\
&\geq 1 - \alpha,
\end{aligned}$$

and, for each $i \in \{0, \ldots, n_0\}$, we have

$$\begin{aligned} P(H_i) &= 1 - P(B^*(a_i, r)) \\ &\geq 1 - \alpha - \epsilon. \end{aligned}$$

Consequently, if we let

$$\mathcal{G}_\epsilon \doteqdot \{G_i \mid i \in \{0, \ldots, n_1\}\} \cup \{H_i \mid i \in \{0, \ldots, n_0\}\},$$

then

$$\inf_{G \in \mathcal{G}_\epsilon} P(G) \geq 1 - \alpha - \epsilon,$$

and from the arbitrariness of ϵ it then follows that

$$\begin{aligned} \delta_w(P, \Delta(X)) &\geq \sup_{\mathcal{G} \in 2^{(\mathcal{T})}, \cap \mathcal{G} = \emptyset} \inf_{G \in \mathcal{G}} P(G) \\ &\geq 1 - \alpha, \end{aligned}$$

and we are done.

□

This result shows that the weak distance is much more informative than the weak topology. Since $\Delta(X)$ is weakly closed, for the weak topology, the distance from any non-Dirac measure to $\Delta(X)$ is infinite, even if that measure is "almost" a Dirac measure, as, for example, the probability measure

$$P \doteqdot (1 - \epsilon) P_x + \epsilon Q,$$

where Q is any probability measure other than P_x, and where ϵ is arbitrarily small. In this case we have

$$\delta_{\mathcal{T}_w}(P, \Delta(X)) = \infty.$$

According to 5.1.12, for the weak distance, however, we have

$$\delta_w(P, \Delta(X)) \leq \epsilon.$$

Some further results for this example are given in the next chapter where we introduce compactness and a measure of compactness.

5.2 FUNCTION SPACES

Let X be a set and let (Y, d) be a metric space. The function space Y^X, consisting of all functions from X to Y, can be endowed with a number of natural uniform and topological structures. A unified way of treating this can be found in Bourbaki [18] and we will follow the approach presented there. Given any collection $\mathfrak{S}$ of subsets of X which covers X and which is closed under the formation of finite unions, the collection of entourages

$$\mathbf{W}(A,\epsilon) \doteqdot \left\{(f,g) \in Y^X \times Y^X \mid \forall x \in A : \ d(f(x), g(x)) < \epsilon\right\},$$

where $A \in \mathfrak{S}$ and $\epsilon > 0$, defines the so-called *uniformity of uniform convergence on $\mathfrak{S}$-sets* or *uniformity of $\mathfrak{S}$-convergence.* We will denote this uniformity by $\mathcal{U}_{\mathfrak{S}}$. The topology derived from this uniformity is called the *topology of uniform convergence on $\mathfrak{S}$-sets* or *topology of $\mathfrak{S}$-convergence.* We will denote this topology by $\mathcal{T}_{\mathfrak{S}}$.

Now, for any subset $A \subset X$, we define the pseudometric

$$d_A : Y^X \times Y^X \longrightarrow \mathbb{R}: (f,g) \longrightarrow \sup_{x \in A} d(f(x), g(x)).$$

Then the set

$$\mathcal{D}_{\mathfrak{S}} \doteqdot \{d_A \mid A \in \mathfrak{S}\},$$

is closed under the formation of finite suprema and thus it is a symmetric basis for a symmetric gauge. The distance generated by this gauge, which, for any $f \in Y^X, \mathcal{A} \subset Y^X$, is given by

$$\delta_{\mathfrak{S}}(f, \mathcal{A}) = \sup_{A \in \mathfrak{S}} \inf_{g \in \mathcal{A}} d_A(f, g),$$

will be called the *distance of $\mathfrak{S}$-convergence.* The associated limit operator will be denoted by $\lambda_{\mathfrak{S}}$ and will be called the *limit operator of $\mathfrak{S}$-convergence* and so on.

In the case where $\mathfrak{S}$ consists of all finite subsets of X, we have several alternative characterizations of $\left(Y^X, \delta_{\mathfrak{S}}\right)$.

5.2.1 Proposition *If $\mathfrak{S}$ consists of all finite subsets of X, then $\left(Y^X, \delta_{\mathfrak{S}}\right)$ is isomorphic to the product space $\left(\prod_{x \in X} Y_x, \prod_{x \in X} \delta_{d_x}\right)$, where, for each $x \in X$, $Y_x = Y$ and $d_x = d$.*

Proof If we set δ the product distance, then it follows from 4.1.1 that, for any $f \in Y^X$ and $\mathcal{A} \subset Y^X$,

$$\begin{aligned}\delta(f,\mathcal{A}) &= \sup_{A\in 2^{(X)}} \inf_{g\in\mathcal{A}} \sup_{y\in A} d\big(\mathrm{pr}_f(y), \mathrm{pr}_g(y)\big) \\ &= \sup_{A\in\mathfrak{S}} \inf_{g\in\mathcal{A}} \sup_{y\in A} d(f(y), g(y)) \\ &= \delta_{\mathfrak{S}}(f,\mathcal{A}).\end{aligned}$$

□

In the following proposition $\mathrm{ev}_x : Y^X \longrightarrow Y$ stands for the *evaluation map* defined by $\mathrm{ev}_x(f) \doteqdot f(x)$, for all $x \in X$ and $f \in Y^X$.

5.2.2 Proposition *If $\mathfrak{S}$ consists of all finite subsets of X, then $\delta_{\mathfrak{S}}$ is initial for the source $\big(\mathrm{ev}_x : Y^X \longrightarrow (Y,\delta_d)\big)_{x\in X}$.*

Proof It follows from 1.10.4 that a basis for the initial gauge is given by

$$\mathcal{H} \doteqdot \left\{ \sup_{x\in X_0} d\circ(\mathrm{ev}_x \times \mathrm{ev}_x) \mid X_0 \in 2^{(X)} \right\},$$

from which the result immediately follows.

□

5.2.3 Proposition *The limit operator $\lambda_{\mathfrak{S}}$ on Y^X is given by the following formula. For any filter Ψ in Y^X and any $f \in Y^X$:*

$$\lambda_{\mathfrak{S}}(\Psi)(f) = \sup_{A\in\mathfrak{S}} \inf_{\mathcal{F}\in\Psi} \sup_{g\in\mathcal{F}} d_A(f,g).$$

Proof This follows from 1.8.31 and the definition of $\mathcal{D}_{\mathfrak{S}}$.

□

5.2.4 Corollary *For any sequence $(f_n)_n$ in Y^X and any $f \in Y^X$:*

$$\lambda_{\mathfrak{S}}(\langle\!\langle (f_n)_n \rangle\!\rangle)(f) = \sup_{A\in\mathfrak{S}} \limsup_{n\to\infty} d_A(f,f_n).$$

□

Again, in the special case where $\mathfrak{S}$ consists of all finite subsets of X, we are able to give an alternative formula for the associated limit operator.

5.2.5 Proposition *If $\mathfrak{S}$ consists of all finite subsets of X, then the limit operator $\lambda_{\mathfrak{S}}$ on Y^X is given by the following formula. For any filter Ψ in Y^X and any $f \in Y^X$:*

$$\lambda_{\mathfrak{S}}(\Psi)(f) = \sup_{x \in X} \inf_{\mathcal{F} \in \Psi} \sup_{g \in \mathcal{F}} d(f(x), g(x)).$$

Proof This follows from 1.10.5 and the fact that $\delta_{\mathfrak{S}}$ is initial for the source $\left(\mathrm{ev}_x : Y^X \longrightarrow (Y, \delta_d)\right)_{x \in X}$ as was shown in 5.2.2.

□

5.2.6 Corollary *If $\mathfrak{S}$ consists of all finite subsets of X, then for any sequence $(f_n)_n$ in Y^X and any $f \in Y^X$:*

$$\lambda_{\mathfrak{S}}(\langle (f_n)_n \rangle)(f) = \sup_{x \in X} \limsup_{n \to \infty} d(f(x), f_n(x)).$$

□

5.2.7 Proposition *The tower $\mathfrak{t}_{\mathfrak{S}}$ on Y^X is given by the following formula. For any $\mathcal{A} \subset Y^X$ and $\epsilon \in \mathbb{R}^+$:*

$$(\mathfrak{t}_{\mathfrak{S}})_\epsilon(\mathcal{A}) = \left\{ f \in Y^X \mid \forall A \in \mathfrak{S}, \forall \theta > \epsilon : \ \mathbf{W}(A, \theta)(f) \cap \mathcal{A} \neq \emptyset \right\}.$$

Proof From the definition of $\delta_{\mathfrak{S}}$ we obtain

$$\begin{aligned} &\delta_{\mathfrak{S}}(f, \mathcal{A}) \leq \epsilon \\ &\Leftrightarrow \forall A \in \mathfrak{S}, \forall \theta > \epsilon, \exists g \in \mathcal{A} : \ d_A(f,g) < \theta \\ &\Leftrightarrow \forall A \in \mathfrak{S}, \forall \theta > \epsilon : \quad \mathbf{W}(A, \theta)(f) \cap \mathcal{A} \neq \emptyset. \end{aligned}$$

□

5.2.8 Proposition *The* **TOP**-*bicoreflection of $(Y^X, \delta_{\mathfrak{S}})$ is $(Y^X, \mathcal{T}_{\mathfrak{S}})$, where $\mathcal{T}_{\mathfrak{S}}$ is the topology of $\mathfrak{S}$-convergence.*

Proof For any $A \in \mathfrak{S}$ and $\epsilon > 0$, let

$$\mathbf{U}_{d_A}(\epsilon) \doteqdot \left\{ (f, g) \in Y^X \times Y^X \mid d_A(f, g) < \epsilon \right\}.$$

Then it suffices to notice that, for any $A \in \mathfrak{S}$, $\epsilon > 0$, and $\theta > \epsilon$, we have

$$\mathbf{U}_{d_A}(\epsilon) \subset \mathbf{W}(A, \epsilon) \subset \mathbf{U}_{d_A}(\theta).$$

Hence, the same inclusions hold for the sections.

□

5.2.9 Proposition *The* pMET$^\infty$*-bicoreflection of* $(Y^X, \delta_{\mathfrak{S}})$ *is* (Y^X, d_∞)*, where* d_∞ *is the uniform metric.*

Proof It suffices to notice that, for any $f, g \in Y^X$, from the fact that $\mathfrak{S}$ is a cover of X, we have

$$\delta_{\mathfrak{S}}(f, \{g\}) = \sup_{A \in \mathfrak{S}} d_A(f, g) = d_\infty(f, g).$$

□

An interesting aspect of function spaces is that under mild conditions the codomain of the functions is usually embedded as a subspace of the function space (topological or uniform). The same result holds here.

5.2.10 Proposition *The map*

$$(Y, \delta_d) \longrightarrow \left(Y^X, \delta_{\mathfrak{S}}\right) : y \longrightarrow \tilde{y},$$

where, for all $x \in X$, $\tilde{y}(x) \doteqdot y$*, is an embedding.*

Proof This follows from the fact that, for any $y \in Y$ and $B \subset Y$,

$$\begin{aligned}\delta_{\mathfrak{S}|\tilde{Y}}\left(\tilde{y}, \tilde{B}\right) &= \sup_{A \in \mathfrak{S}} \inf_{z \in B} d_A(\tilde{y}, \tilde{z}) \\ &= \sup_{A \in \mathfrak{S}} \inf_{z \in B} d(y, z) \\ &= \inf_{z \in B} d(y, z) \\ &= \delta_d(y, B).\end{aligned}$$

□

5.2.11 Example If we take for $\mathfrak{S}$ the collection of all finite subsets of X, then the **TOP**-bicoreflection of $(Y^X, \delta_{\mathfrak{S}})$ is the topology of pointwise convergence or product topology. This topology is not metrizable in general but, as we have seen, it can be "distancized" in a natural way. To demonstrate what kind of numerical values one obtains with this distance let us consider the following simple situation. Let $X = Y \doteqdot \mathbb{R}$. Further let

$$\mathcal{A} \doteqdot \left\{g \in Y^X \mid \{g \neq 0\} \text{ finite}\right\}$$

and

$$f : X \longrightarrow Y : x \longrightarrow 1.$$

Since we have that $f \in \overline{\mathcal{A}}$ for the product topology we also have that $\delta_{\mathfrak{S}}(f, \mathcal{A}) = 0$. Let us change f slightly in the following way:

$$f_\epsilon : X \longrightarrow Y : x \longrightarrow \begin{cases} 1 & x \neq 0, \\ 1 + \epsilon & x = 0. \end{cases}$$

Then $f_\epsilon \notin \overline{\mathcal{A}}$ and hence $\delta_{\mathfrak{S}}(f_\epsilon, \mathcal{A}) \neq 0$; however, unlike the situation for the product topology, the value does not "jump to ∞". Actually we have

$$\delta_{\mathfrak{S}}(f_\epsilon, \mathcal{A}) = \sup_{A \in 2^{(X)}} \inf_{g \in \mathcal{A}} \sup_{a \in A} |f_\epsilon(a) - g(a)| = \epsilon.$$

In the next chapter, where we study the completeness of uniform approach spaces, we will return to this example.

5.3 HYPERSPACES

In the literature a large variety of topologies, uniformities, and metrics have been considered on hyperspaces of topological, uniform, or metric spaces. See e.g. Beer and Luchetti [11], Beer [9], and Beer [10]. In particular in the case of metric spaces, the so-called *Wijsman topology* on the hyperspace of all closed sets has been extensively studied.

Let (X, d) be a metric space. We denote by $CL(X)$ the set of all nonempty closed subsets of X.

We begin by recalling the definitions of the Hausdorff metric and of the Wijsman topology.

The *Hausdorff metric* on $CL(X)$ can be defined in several equivalent ways, one of which is important for us. We denote this metric by h_d. It is given by the following formula (see e.g. Beer [10]). For any $A, B \in CL(X)$:

$$h_d(A, B) = \sup_{x \in X} |\delta_d(x, A) - \delta_d(x, B)|.$$

The Wijsman topology too can be introduced in several equivalent ways. Two of these are important for our considerations.

First, it is the initial topology on $CL(X)$ for the source

$$\left(CL(X) \longrightarrow \mathbb{R}^+ : A \longrightarrow \delta_d(x,A)\right)_{x\in X},$$

where $\mathbb{R}^+$ is equipped with the usual Euclidean topology.

Second, it can also be characterized as being the initial topology for the source

$$CL(X) \longrightarrow \left(\mathbb{R}^+\right)^X : A \longrightarrow \delta_d(\cdot, A),$$

where $\mathbb{R}^+$ is again equipped with the usual Euclidean topology and $\left(\mathbb{R}^+\right)^X$ is equipped with the product topology. In both cases the usual topology on the real line plays a crucial role and the Wijsman topology owes its existence to the fact that we are able to construct initial topologies. In the first method we are simply constructing the Wijsman topology as an initial topology of the usual topology on $\mathbb{R}^+$ for a collection of maps, and in the second method we are actually identifying $CL(X)$ with the set of all distance functionals $\{\delta_d(\cdot, A) \mid A \in CL(X)\}$ and we construct the Wijsman topology as a subspace topology of a product (or pointwise) topology.

If, however, we start not with the usual topology on $\mathbb{R}^+$ but with the usual metric, then classically we are unable to perform either of these constructions. Again, using approach spaces we can overcome these obstacles.

For each finite subset F of X we define

$$d_F : CL(X) \times CL(X) \longrightarrow [0, \infty]$$

by

$$d_F(A, B) \doteq \sup_{x\in F} |\delta_d(x, A) - \delta_d(x, B)|.$$

Now consider the collection

$$\mathcal{D}_{W_d} \doteq \left\{d_F \mid F \in 2^{(X)}\right\}.$$

The set $\mathcal{D}_{W_d}$ of pseudometrics is closed under the formation of finite suprema and hence is a symmetric basis for a symmetric gauge. The distance generated by this gauge is given by

$$\delta_{W_d} : CL(X) \times 2^{CL(X)} \longrightarrow [0, \infty] : (A, \mathcal{A}) \longrightarrow \sup_{d\in\mathcal{D}_{W_d}} \inf_{B\in\mathcal{A}} \inf_{a\in A} \inf_{b\in B} d(a, b).$$

We will refer to this distance as the *Wijsman distance*, and our first task of course is to justify this terminology.

5.3.1 Theorem *The topological bicoreflection of* $(CL(X), \delta_{W_d})$ *is* $CL(X)$ *equipped with the Wijsman topology* $\mathcal{T}_{W_d}$.

Proof It follows from 4.3.3 that a basis for the neighborhoods of $A \in CL(X)$ in the underlying topology of $(CL(X), \delta_{W_d})$ is given by the collection

$$\{B \in CL(X) \mid d_F(A, B) < \epsilon\} \qquad \forall F \in 2^{(X)}, \ \forall \epsilon > 0.$$

This, however, is precisely a basis for the neighborhoods of A in the Wijsman topology $\mathcal{T}_{W_d}$.

□

5.3.2 Theorem *The* ∞*p-metric bicoreflection of* $(CL(X), \delta_{W_d})$ *is* $CL(X)$ *equipped with the Hausdorff metric* h_d.

Proof It follows from 4.3.3 that, for all $A, B \in CL(X)$, we have

$$\begin{aligned} d_{\delta_{W_d}}(A, B) &= \sup_{F \in 2^{(X)}} d_F(A, B) \\ &= \sup_{x \in X} d_{\{x\}}(A, B) \\ &= h_d(A, B). \end{aligned}$$

□

In Lechicki and Levi [64] it was shown that the Wijsman topology on $CL(X)$ is metrizable if and only if $(X, \mathcal{T}_d)$ is separable. Such a condition is not required in our case. $(CL(X), \mathcal{T}_{W_d})$ is always canonically "distancizable" by δ_{W_d}. The canonicity is enhanced by the following results which are to be compared with the analogous results for the Wijsman topology in the setting of the construct of topological spaces. Notice that the difference with the case for the Wijsman topology has two aspects: (1) we do not work in **TOP** but in **AP**, (2) we do not equip the real line with the Euclidean topology but with the Euclidean metric.

5.3.3 Proposition *The Wijsman distance is the initial structure on* $CL(X)$ *for the source (in* **AP***)*

$$\left(CL(X) \longrightarrow \mathbb{R}^+ : A \longrightarrow \delta_d(x, A)\right)_{x \in X},$$

where $\mathbb{R}^+$ *is equipped with the usual Euclidean metric.*

Proof As before, we denote the Euclidean metric by d_E. According to 1.10.4 the initial approach structure is given by the basis for the gauge

$$\left\{\sup_{x\in F} d_E \circ (d(x,\cdot) \times d(x,\cdot)) \mid F \in 2^{(X)}\right\}.$$

The result then follows from the fact that, for all $F \in 2^{(X)}$ and all $A, B \in CL(X)$, we have

$$\begin{aligned}\sup_{x\in F} d_E(\delta_d(x,A),\delta_d(x,B)) &= \sup_{x\in F} |\delta_d(x,A) - \delta_d(x,B)|,\\ &= d_F(A,B).\end{aligned}$$

□

5.3.4 Proposition *The Wijsman distance is the initial structure on $CL(X)$ for the source (in* **AP***)*

$$CL(X) \longrightarrow (\mathbb{R}^+)^X : A \longrightarrow \delta_d(\cdot, A),$$

where $\mathbb{R}^+$ is again equipped with the usual Euclidean metric and $(\mathbb{R}^+)^X$ is equipped with the product distance (in **AP***).*

Proof The product distance on $(\mathbb{R}^+)^X$ is given by

$$\delta(f,\mathcal{F}) \doteqdot \sup_{F\in 2^{(X)}} \inf_{g\in\mathcal{F}} \sup_{x\in F} |f(x) - g(x)| \qquad f \in (\mathbb{R}^+)^X,\ \mathcal{F} \subset (\mathbb{R}^+)^X$$

Consequently, the initial distance is given by

$$\begin{aligned}\delta'(A,\mathcal{A}) &= \sup_{F\in 2^{(X)}} \inf_{B\in\mathcal{A}} \sup_{x\in F} |\delta_d(x,A) - \delta_d(x,B)|\\ &= \sup_{F\in 2^{(X)}} \inf_{B\in\mathcal{A}} d_F(A,B),\end{aligned}$$

for all $A \in CL(X)$ and $\mathcal{A} \subset CL(X)$, which is precisely the Wijsman distance.

□

A topological (or uniform structure) on a hyperspace is called admissible, Michael [88], if the map $x \longrightarrow \{x\}$ is well defined and an embedding. Analogously we will say that a distance δ on $CL(X)$ is *admissible* if

$$(X,\delta_d) \longrightarrow (CL(X),\delta) : x \longrightarrow \{x\}$$

is well defined and an embedding.

5.3.5 Proposition *The Wijsman distance on $CL(X)$ is admissible.*

Proof Since $(X, \mathcal{T}_d)$ is Hausdorff the function

$$\psi : (X, \delta_d) \longrightarrow (CL(X), \delta_{W_d}) : x \longrightarrow \{x\}$$

is well defined. Now if $x \in X$ and $A \subset X$, then on the one hand we have

$$\begin{aligned}\delta_{W_d}(\psi(x), \psi(A)) &= \sup_{F \in 2^{(X)}} \inf_{a \in A} \sup_{y \in F} |d(y, x) - d(y, a)| \\ &\geq \inf_{a \in A} |d(x, x) - d(x, a)| \\ &= \delta_d(x, A),\end{aligned}$$

whereas on the other hand we have

$$\begin{aligned}\delta_{W_d}(\psi(x), \psi(A)) &= \sup_{F \in 2^{(X)}} \inf_{a \in A} \sup_{y \in F} |d(y, x) - d(y, a)| \\ &\leq \sup_{F \in 2^{(X)}} \inf_{a \in A} \sup_{y \in F} d(x, a) \\ &= \delta_d(x, A).\end{aligned}$$

Consequently, ψ is an embedding.

□

Identifying a set with the set of its singletons and a metric space with its "copy" in **AP**, the foregoing result says that (X, d) is a metric subspace of $(CL(X), \delta_{W_d})$, or that $\delta_{W_d}|_{X \times X} = d$. Hence, the link between metric and Wijsman distance is much stronger than between metric and Wijsman topology, since under certain conditions, described in Costantini *et al.* [26], different metrics on X may generate the same Wijsman topology. The foregoing result actually shows that $\mathbf{MET}(X) \longrightarrow \mathbf{AP}(X) : d \longrightarrow \delta_{W_d}$ is an injection.

5.3.6 Theorem *If (X, d) is separable and Y is a countable dense subset of X, then δ_{W_d} is generated by the countable set of pseudometrics*

$$\mathcal{D}_0 \doteqdot \left\{ d_F \mid F \in 2^{(Y)} \right\}.$$

Proof Evidently, for any $A \in CL(X)$ and $\mathcal{A} \subset CL(X)$, we have

$$\delta_{W_d}(A, \mathcal{A}) \geq \sup_{F \in 2^{(Y)}} \inf_{B \in \mathcal{B}} d_F(A, B).$$

Conversely, let $\epsilon > 0$ and let $F \in 2^{(X)}$. Then, for each $x \in F$, there exists $y_x \in Y$ such that $d(x, y_x) \leq \frac{\epsilon}{2}$. It follows that if we let $F_\epsilon \doteqdot \{y_x \mid x \in F\}$, then

$$\begin{aligned}\delta_{W_d}(A, \mathcal{A}) &= \sup_{F \in 2^{(X)}} \inf_{B \in \mathcal{B}} d_F(A, B) \\ &\leq \sup_{F \in 2^{(X)}} \inf_{B \in \mathcal{B}} d_{F_\epsilon}(A, B) + \epsilon \\ &\leq \sup_{F \in 2^{(Y)}} \inf_{B \in \mathcal{B}} d_F(A, B) + \epsilon,\end{aligned}$$

which by the arbitrariness of ϵ proves our claim.

□

It is well known that if the metric space (X, d) is totally bounded then the Wijsman topology is metrized by the Hausdorff metric. A last interesting result concerning the relationship between the Hausdorff metric and the Wijsman distance considerably strengthens this property.

5.3.7 Theorem *If (X, d) is totally bounded, then $\delta_{W_d} = \delta_{h_d}$.*

Proof Since we have that $\delta_{W_d} \leq \delta_{h_d}$ it suffices to prove the other inequality. For this let $A \in CL(X)$ and let $\mathcal{A} \subset CL(X)$. Further let $\epsilon > 0$ be fixed, and let $F_0 \subset X$ be a finite set such that

$$\bigcup_{x \in F_0} B\left(x, \frac{\epsilon}{2}\right) = X.$$

Then, for any $y \in X$, there exists $x \in F_0$ such that, for all $B \in \mathcal{A}$,

$$|d(y, A) - d(y, B)| \leq |d(x, A) - d(x, B)| + \epsilon,$$

and consequently

$$\begin{aligned}\delta_{h_d}(A, \mathcal{A}) &= \inf_{B \in \mathcal{A}} \sup_{x \in X} |d(x, A) - d(x, B)| \\ &\leq \sup_{F \in 2^{(X)}} \inf_{B \in \mathcal{A}} \sup_{x \in F} |d(x, A) - d(x, B)| + \epsilon \\ &= \delta_{W_d}(A, \mathcal{A}) + \epsilon,\end{aligned}$$

which, by the arbitrariness of ϵ, proves our claim.

□

In the next chapter, where we study the completeness of uniform approach spaces, we will return to hyperspaces.

5.4 PROBABILISTIC METRIC SPACES

Let $(X, \mathfrak{F}, T_m)$ be a so-called T_m-Menger space, i.e. $\mathfrak{F}$ is a probabilistic metric on X with respect to the triangular norm T_m. We refer to Appendix B or Schweizer and Sklar [106] for more detailed information on these concepts.

For each $a > 0$, we define the function

$$\varphi^a : X \times X \longrightarrow [0, \infty] : (x, y) \longrightarrow 1 - \mathfrak{F}(x, y)(a).$$

5.4.1 Proposition

For any $a, b > 0$ and $x, y, z \in X$, we have

$$\varphi^{a+b}(x, z) \leq \varphi^a(x, y) + \varphi^b(y, z).$$

Proof This follows from

$$\begin{aligned} \varphi^{a+b}(x, z) &= 1 - \mathfrak{F}(x, z)(a + b) \\ &\leq 1 - T_m(\mathfrak{F}(x, y)(a), \mathfrak{F}(y, z)(b)) \\ &\leq 1 - (\mathfrak{F}(x, y)(a) + \mathfrak{F}(y, z)(b) - 1) \\ &= \varphi^a(x, y) + \varphi^b(y, z). \end{aligned}$$

□

From this proposition it follows that if we let

$$\mathcal{B}_{\mathfrak{F}}(x) \doteqdot \{\varphi^a(x, \cdot) \mid a > 0\} \qquad \forall x \in X,$$

then $\left(\mathcal{A}_{\mathfrak{F}}(x) \doteqdot \hat{\mathcal{B}}_{\mathfrak{F}}(x)\right)_{x \in X}$ is an approach system on X. We call this the *approach system underlying* or *generated by the probabilistic metric* $\mathfrak{F}$. The other approach structures on X determined by this approach system will also be denoted by means of the subscript $\mathfrak{F}$. Thus the associated distance is denoted by $\delta_{\mathfrak{F}}$, and so on.

Now for each $a > 0$ we define the function $d^a : X \times X \longrightarrow [0, \infty]$ determined by

$$d^a(x, y) \doteqdot \inf \left\{\theta \in]0, 1] \mid \varphi^{\theta a}(x, y) \leq \theta\right\} \qquad \forall x, y \in X.$$

5.4.2 Proposition *For any $a > 0$, the function $d^a : X \times X \longrightarrow [0,\infty]$ is a p-metric.*

Proof The only property which needs verification is the triangular inequality, but this follows immediately from 5.4.1.

□

5.4.3 Proposition *The distance $\delta_{\mathfrak{F}}$ is generated by the set of p-metrics $\{d^a \mid a > 0\}$.*

Proof Let δ stand for the distance generated by the set $\{d^a \mid a > 0\}$. Fix $x \in X$ and $A \subset X$. If $\delta_{\mathfrak{F}}(x, A) < \theta$, then there exists $y \in A$ such that $\varphi^{\theta a}(x, y) < \theta$ and hence $d^a(x, y) \leq \theta$ which proves that $\delta(x, A) \leq \delta_{\mathfrak{F}}(x, A)$. Conversely, if $0 < \theta < \delta_{\mathfrak{F}}(x, A)$, then there exists $a > 0$ such that $\theta < \inf_{y \in A} \varphi^a(x, y)$. Letting $b \doteqdot a\theta^{-1}$ it follows that $\theta \leq \inf_{y \in A} d^b(x, y)$ which proves that $\delta_{\mathfrak{F}}(x, A) \leq \delta(x, A)$.

□

5.4.4 Proposition *The limit operator $\lambda_{\mathfrak{F}}$ on X is given by the following formulas. For any filter $\mathcal{F}$ in X and any $x \in X$:*

$$\begin{aligned}\lambda_{\mathfrak{F}}(\mathcal{F})(x) &= \sup_{a>0} \inf_{G\in\mathcal{F}} \sup_{y\in G} \varphi^a(x, y)\\ &= \sup_{a>0} \inf_{G\in\mathcal{F}} \sup_{y\in G} d^a(x, y).\end{aligned}$$

Proof This follows from 1.8.30 and the definition of $\delta_{\mathfrak{F}}$.

□

5.4.5 Proposition *The tower $\mathfrak{t}_{\mathfrak{F}}$ on X is given by the following formulas. For any $A \subset X$ and $\epsilon \in \mathbb{R}^+$:*

$$\begin{aligned}(\mathfrak{t}_{\mathfrak{F}})_\epsilon(A) &= \{x \in X \mid \forall a > 0,\ \forall\theta > \epsilon\ \exists y \in A:\ \varphi^a(x,y) < \theta\}\\ &= \{x \in X \mid \forall a > 0,\ \forall\theta > \epsilon : B_{d^a}(x,\theta) \cap A \neq \emptyset\}.\end{aligned}$$

Proof This follows from 1.8.19 and the definition of $\delta_{\mathfrak{F}}$.

□

5.4.6 Proposition *The* **TOP**-*bicoreflection of* $(X, \delta_{\mathfrak{F}})$ *is* $(X, \mathcal{T}_{\mathfrak{F}})$, *where* $\mathcal{T}_{\mathfrak{F}}$ *is the strong topology.*

Proof For any $a > 0$ and $\epsilon > 0$ let

$$\begin{aligned} \mathbf{U}_{d^a}(\epsilon) &\doteqdot \{(x,y) \in X \times X \mid d^a(x,y) < \epsilon\} \\ &= \{(x,y) \mid \exists \theta < \epsilon : \ \mathfrak{F}(x,y)(\theta a) \geq 1 - \theta\}. \end{aligned}$$

Then it suffices to notice that, for any $a \in]0,1[$, $\epsilon > 0$, and $\theta < \epsilon$, we have

$$\mathbf{U}_{d_a}(\epsilon) \subset \mathbf{U}^{\epsilon}_{\mathfrak{F}} \text{ and } \mathbf{U}^{\theta a}_{\mathfrak{F}} \subset \mathbf{U}_{d_a}(\epsilon),$$

where $\mathbf{U}^{\epsilon}_{\mathfrak{F}}$ stands for the strong ϵ-entourage as introduced in Appendix B. Hence the same inclusions hold for the sections.

□

5.4.7 Proposition

The p**MET**$^\infty$-*bicoreflection of* $(X, \delta_{\mathfrak{F}})$ *is* $(X, d_{\mathfrak{F}})$, *where* $d_{\mathfrak{F}}$ *is the metric determined by*

$$d_{\mathfrak{F}}(x,y) \doteqdot 1 - \mathfrak{F}(x,y)(0+) \qquad \forall x,y \in X.$$

Proof It suffices to notice that, for any $x, y \in X$,

$$\begin{aligned} d_{\mathfrak{F}}(x,y) &= \sup_{a>0} \varphi^a(x,y) \\ &= \sup_{a>0} (1 - \mathfrak{F}(x,y)(a)), \end{aligned}$$

and to remember that $\mathfrak{F}(x,y)$ is increasing.

□

In Schweizer and Sklar [106] two points of a probabilistic metric space were called *perhaps indiscernible* if $\mathfrak{F}(x,y)(0+) > 0$, in other words if $d_{\mathfrak{F}}(x,y) < 1$. Hence we can refer to this ∞p-metric as the ∞p*-metric of indiscernibility*. It is interesting to see that the ∞p-metric bicoreflection again seems to measure some separatedness of the space. This should be compared to our remarks concerning the ∞p-metric and the ∞pq-metric bicoreflections of a topological approach space in the second chapter.

5.5 SPACES OF RANDOM VARIABLES

A special case of the foregoing example, but more than sufficiently important in its own right, is the following example. Let $(\Omega, \mathcal{A}, P)$ be a probability space, (M, d) a separable metric space, and let $\mathcal{R}(M)$ be the set of all M-valued random variables on Ω. Sherwood [110] proved that this space is a T_m-Menger space under the probabilistic metric

$$\mathfrak{F}(X,Y)(a) \doteqdot P(\{\omega \in \Omega \mid d(X(\omega),Y(\omega)) < a\}).$$

By the foregoing example this space is then naturally endowed with an approach structure, and using the same notations we see that, in the present case, the functions φ^a, $a > 0$, are determined by

$$\varphi^a(X,Y) = P(\{\omega \in \Omega \mid d(X(\omega),Y(\omega)) \geq a\}) \qquad X, Y \in \mathcal{R}(M).$$

We will denote the distance generated by this collection by $\delta_{\mathcal{R}(M)}$.

5.5.1 Proposition *The* **TOP**-*bicoreflection of* $(\mathcal{R}(M), \delta_{\mathcal{R}(M)})$ *is* $(\mathcal{R}(M), \mathcal{T}_{\mathcal{R}(M)})$, *where* $\mathcal{T}_{\mathcal{R}(M)}$ *is the topology of convergence in probability.*

Proof It suffices to notice that, for any $X \in \mathcal{R}(M)$, a basis for the neighborhoods of X in the **TOP**-bicoreflection of $(\mathcal{R}(M), \delta_{\mathcal{R}(M)})$ is given by the collection of sets

$$\{Y \in \mathcal{R}(M) \mid P(\{\omega \in \Omega \mid d(X(\omega),Y(\omega)) \geq a\}) < \epsilon\} \qquad a \in [0,\infty[, \epsilon > 0,$$

and that this collection is also precisely a basis for the topology of convergence in probability.

□

5.5.2 Proposition *The* **pMET**$^\infty$-*bicoreflection of* $(\mathcal{R}(M), \delta_{\mathcal{R}(M)})$ *is* $(\mathcal{R}(M), d_{\mathcal{R}(M)})$, *where* $d_{\mathcal{R}(M)}$ *is the* ∞*p-metric determined by*

$$d_{\mathcal{R}(M)}(X,Y) = P(\{\omega \in \Omega \mid d(X(\omega),Y(\omega)) > 0\}).$$

Proof It suffices to notice that, for any $X, Y \in \mathcal{R}(M)$, we have

$$\begin{aligned} d_{\mathcal{R}(M)}(X,Y) &= 1 - \mathfrak{F}(X,Y)(0+) \\ &= \lim_{a\downarrow 0} P(\{\omega \in \Omega \mid d(X(\omega), Y(\omega)) \geq a\}) \\ &= P(\{\omega \in \Omega \mid d(X(\omega), Y(\omega)) > 0\}). \end{aligned}$$

□

This ∞p-metric attains the value 0 if and only if the random variables X and Y are equal almost everywhere; hence we can refer to it as the *∞p-metric of equality almost everywhere.*

5.5.3 Proposition *The limit operator $\lambda_{\mathcal{R}(M)}$ on $\mathcal{R}(M)$ (for sequences of random variables) is given by the following formula:*

$$\lambda_{\mathcal{R}(M)}\langle (X_n)_n \rangle (X) = \sup_{a>0} \limsup_{n\to\infty} P(\{\omega \in \Omega \mid d(X(\omega), X_n(\omega)) \geq a\}).$$

Proof This follows at once from 5.4.4.

□

We end this example with a result linking the structure introduced for random variables with the one introduced for probability measures. We recall that if X is a random variable, then its *law* is the probability measure P_X in $\mathcal{P}$ (where $\mathcal{P}$ stands for the set of probability measures on M) defined by $P_X(B) \doteqdot P\big(X^{-1}(B)\big)$, for all $B \in \mathcal{B}$ (where $\mathcal{B}$ stands for the Borel sets in M).

It is well known that convergence in probability of a sequence of random variables implies weak convergence of their laws. Since $\delta_{\mathcal{R}(M)}$ has the topology of convergence in probability as $\mathbf{TOP}$-bicoreflection and δ_w has the weak topology as $\mathbf{TOP}$-bicoreflection it is natural to ask what becomes of the above property in the new setting. The following result gives an answer to that question.

5.5.4 Proposition *The function*

$$L : \big(\mathcal{R}(M), \delta_{\mathcal{R}(M)}\big) \longrightarrow (\mathcal{P}, \delta_w) : X \longrightarrow P_X$$

is a contraction.

Proof Let $X \in \mathcal{R}(M)$ and $\mathcal{A} \subset \mathcal{R}(M)$. Then we have to show that $\delta_w(L(X), L(\mathcal{A})) \leq \delta_{\mathcal{R}(M)}(X, \mathcal{A})$, where

$$\delta_w(L(X), L(\mathcal{A})) = \sup_{\mathcal{H} \in 2^{(\mathcal{U}(X,[0,1]))}} \inf_{Y \in \mathcal{A}} d_3^{\mathcal{H}}(L(X), L(Y))$$

and

$$\delta_{\mathcal{R}(M)}(X, \mathcal{A}) = \sup_{a>0} \inf_{Y \in \mathcal{A}} \varphi^a(X, Y).$$

Let $\mathcal{H} \in 2^{(\mathcal{U}(X,[0,1]))}$ and $\epsilon > 0$ be fixed. For all $f \in \mathcal{H}$, choose $\theta_f > 0$ such that, for all $x, y \in M$, $d(x, y) \leq \theta_f \Rightarrow |f(x) - f(y)| \leq \epsilon$ and put $\theta \doteqdot \min_{f \in \mathcal{H}} \theta_f$. Then it will suffice to prove that, for all $Y \in \mathcal{A}$, we have $d_3^{\mathcal{H}}(L(X), L(Y)) \leq \varphi^{\theta}(X, Y) + \epsilon$.

$$\begin{aligned}
d_3^{\mathcal{H}}(L(X), L(Y)) &= \sup_{f \in \mathcal{H}} \left| \int f \circ X \mathrm{d}P - \int f \circ Y \mathrm{d}P \right| \\
&\leq \sup_{f \in \mathcal{H}} \Bigg(\int_{\{d(X,Y)<\theta\}} |f \circ X - f \circ Y| \mathrm{d}P \\
&\qquad + \int_{\{d(X,Y)\geq\theta\}} |f \circ X - f \circ Y| \mathrm{d}P \Bigg) \\
&\leq \sup_{f \in \mathcal{H}} (\epsilon + P(\{d(X, Y) \geq \theta\})) \\
&= \epsilon + \varphi^{\theta}(X, Y).
\end{aligned}$$

□

5.5.5 Corollary *For any sequence of random variables $(X_n)_n$ and any random variable X, we have $\lambda_w\langle(L(X_n))_n\rangle(L(X)) \leq \lambda_{\mathcal{R}(M)}\langle(X_n)_n\rangle(X)$.*

□

5.5.6 Corollary *If a sequence of random variables $(X_n)_n$ converges in probability to a random variable X, then it also converges in law to X.*

□

5.5.7 Corollary *If a sequence of random variables $(X_n)_n$ converges to a random variable X in the topology generated by the metric of equality almost everywhere, then their laws converge to the law of X in the $\mathbf{L}^1$-metric.*

□

A converse result to 5.5.6 also holds in case the limit random variable is constant. This result too has a nice generalization.

5.5.8 Proposition *If $x \in M$ and $(X_n)_n$ is a sequence of random variables, then we have that $\lambda_{\mathcal{R}(M)}\langle (X_n)_n\rangle(x) \leq \lambda_w\langle (L(X_n))_n\rangle(P_x)$.*

Proof Suppose that $0 < a < \lambda_{\mathcal{R}(M)}\langle (X_n)_n\rangle(x)$; then it follows from 5.5.3 that we can find $b > 0$ such that

$$\forall n\, \exists m \geq n : \ P(\{d(X_m, x) \geq b\}) \geq a.$$

Now define

$$f : M \longrightarrow [0,1] : y \longrightarrow \frac{d(x,y)}{b} \wedge 1.$$

Then $f \in C(M, [0,1])$ and

$$\begin{aligned}\left|\int f\mathrm{d}P_{X_m} - \int f\mathrm{d}P_x\right| &= \left|\int f\mathrm{d}P_{X_m}\right| \\ &\geq \left|\int_{\{d(X_m,x)\geq b\}} f\circ X_m \mathrm{d}P\right| \\ &\geq a,\end{aligned}$$

and consequently $a \leq \lambda_w\langle((L(X_n))_n)\rangle(P_x)$, which proves our claim.

□

5.5.9 Corollary *If $x \in M$ and $(X_n)_n$ is a sequence of random variables which converges in law to P_x, then it also converges in probability to the random variable with constant value x.*

□

To conclude this chapter of examples we give a list of the structures which we have seen together with their topological and metric coreflections.

Basic space	**TOP-coreflection**	**pMET$^\infty$-coreflection**
Space of measures on a metric space (5.1)	Weak topology see e.g. Billingsley [13] (5.1.7)	L^1-metric see e.g. Parthasarathy and Steerneman [98] (5.1.9)
Function space with metric range $\mathfrak{S}=$ finite sets (5.2)	Pointwise topology see e.g. Bourbaki [18] (5.2.8)	Uniform metric see e.g. Bourbaki [18] (5.2.9)
Function space with metric range $\mathfrak{S}=$ compact sets (5.2)	Uniform convergence on compact sets see e.g. Bourbaki [18] (5.2.8)	Uniform metric see e.g. Bourbaki [18] (5.2.9)
Hyperspace of a metric space (5.3)	Wijsman topology see e.g. Lechicki and Levi [64] (5.3.1)	Hausdorff distance see e.g. Michael [88] (5.3.2)
Probabilistic metric space (5.4)	Strong topology see e.g. Schweizer and Sklar [106] (5.4.6)	Metric of indiscernibility see e.g. Schweizer and Sklar [106] (5.4.7)
Space of metric-valued random variables (5.5)	Topology of convergence in probability see e.g. Billingsley [13] (5.5.1)	Metric of equality almost everywhere see e.g. Billingsley [13] (5.5.2)

6
APPROACH PROPERTIES

In this chapter we will study some natural invariants or properties of approach spaces. In view of the fact that approach spaces generalize at the same time topological spaces and metric spaces, the properties which we study will be generalizations of similar topological or metric properties. As a general rule we will adopt the following convention. If (P) is a topological property then we will say that an approach space has (P) if and only if its topological coreflection has (P). This rule will not be followed for properties of metric spaces, however.

Before we start with the formal definitions and results we would like to reflect on the precise nature of approach spaces and the possibilities for defining natural invariants. If a property is a topological invariant, then it is actually a property on the equivalence classes of homeomorphic spaces. Obviously, and at first sight somewhat uninterestingly, the topology itself is a topological invariant. In the same vein the derived convergence structure, the closure operator, and the neighborhood system are topological invariants. We have analogs of all these systems in approach spaces, and what should strike us about these analogs is that they all quantify in some way their topological counterparts. The distance quantifies the closure operator, the limit operator quantifies the convergence structure, and the local distances quantify the neighborhood system.

This opens up an entirely new way of thinking about invariants and properties in approach spaces. Rather than being restricted to asserting that a space does or does not have a certain property we should now have at our disposal machinery by means of which we can define measures of properties, at least in so far that we are dealing with generalizations of properties in **TOP**. This idea is not new. In [62] Kuratowski introduced what he called a measure of noncompactness in the setting of complete metric spaces. A purely topological concept was adapted in a numerical way to be used in the setting of metric spaces in order to measure the deviation an object may have from being compact.

In what follows we will develop some such measures of properties in a systematic way, restricting ourselves, however, to quantifications of two of the most basic topological properties, namely compactness and connectedness.

Since approach spaces also generalize metric spaces we also discuss a notion of completeness in our setting, and this we do in the last section.

6.1 COMPACTNESS

A particularly interesting variant of Kuratowski's original measure of noncompactness is the so-called Hausdorff or ball measure of noncompactness. It is defined as follows.

Suppose that (X, d) is a p-metric space and that $A \subset X$; then

$$m_H(A) \doteqdot \inf \left\{ \epsilon \in \mathbb{R}^+ \mid \exists x_1, \ldots, x_n \in X : \; A \subset \bigcup_{i=1}^{n} B(x_i, \epsilon) \right\}$$

is called the *Hausdorff measure of noncompactness*. It differs from the original measure introduced by Kuratowski only slightly.

The value

$$m_K(A) \doteqdot \inf \left\{ \epsilon \in \mathbb{R}^+ \mid \exists X_1, \ldots, X_n \subset X : \max_{i=1}^{n} diam(X_i) \leq \epsilon, \; A \subset \bigcup_{i=1}^{n} X_i \right\}$$

is called the *Kuratowski measure of noncompactness*. It is easily seen that for any $A \subset X$ we have $m_H(A) \leq m_K(A) \leq 2m_H(A)$.

A first remarkable observation is that both measures are metric dependent, whereas compactness is only dependent on the topology underlying the metric. This rather confusing state of affairs is cleared up if we reconsider the ideas behind these measures in the setting of approach spaces. In this section we show that the Hausdorff measure of noncompactness arises as a very canonical concept in the setting of approach spaces. As far as terminology is concerned, and in spite of the fact that this may cause some confusion, we prefer to call this a measure of compactness.

6.1.1 Definition Given an approach space (X, δ), we define the *measure of compactness* of X as

$$\mu_c(X) \doteqdot \sup_{\mathcal{F} \in \mathbf{F}(X)} \inf_{x \in X} \alpha \mathcal{F}(x).$$

The idea behind this definition is the following. Compactness means every filter should have an adherence point. Therefore the information given by μ_c is based on the verification for all filters of what their "best" adherence points are. Before continuing we give a number of equivalent forms of this definition.

6.1.2 Proposition *For any approach space (X, δ), we have*

$$\begin{aligned}\mu_c(X) &= \sup_{\mathcal{U}\in \mathbf{U}(X)} \inf_{x\in X} \alpha\mathcal{U}(x)\\ &= \sup_{\mathcal{U}\in \mathbf{U}(X)} \inf_{x\in X} \lambda\mathcal{U}(x)\\ &= \sup_{\varphi\in \prod_{x\in X} \mathcal{B}(x)} \inf_{Y\in 2^{(X)}} \sup_{z\in X} \inf_{x\in Y} \varphi(x)(z),\end{aligned}$$

where $(\mathcal{B}(x))_{x\in X}$ is a basis for the approach system.

Proof Let us denote the three expressions of the proposition (in order) by c_1, c_2, and c_3. Then it follows from 1.8.36 that $c_1 = c_2$ and further it is obvious that we have $\mu_c(X) \geq c_1$. To prove that $c_3 \geq \mu_c(X)$, suppose that $\mu_c(X) > r$. Then by 1.8.30 and 1.8.36 we can find an ultrafilter $\mathcal{U}$ such that

$$\begin{aligned}\sup_{\psi\in \prod_{x\in X} \mathcal{B}(x)} \inf_{x\in X} \inf_{U\in\mathcal{U}} \sup_{y\in U} \psi(x)(y) &= \inf_{x\in X} \sup_{\psi\in\mathcal{B}(x)} \inf_{U\in\mathcal{U}} \sup_{y\in U} \psi(y)\\ &= \inf_{x\in X} \alpha\mathcal{U}(x)\\ &> r.\end{aligned}$$

Consequently, there exists $\psi \in \prod_{x\in X} \mathcal{B}(x)$ such that, for all $x \in X$,

$$\inf_{U\in\mathcal{U}} \sup_{y\in U} \psi(x)(y) > r.$$

Now suppose that for some $Y \in 2^{(X)}$ we have

$$\sup_{z\in X} \inf_{x\in Y} \psi(x)(z) \leq r.$$

Then the collection $\{\{\psi(x) \leq r\} \mid x \in Y\}$ is a finite cover of X and thus there exists some $x_0 \in Y$ such that $\{\psi(x_0) \leq r\} \in \mathcal{U}$. Then, however, we find that

$$\begin{aligned}\inf_{U\in\mathcal{U}} \sup_{y\in U} \psi(x_0)(y) &\leq \sup_{y\in\{\psi(x_0)\leq r\}} \psi(x_0)(y)\\ &\leq r,\end{aligned}$$

which is a contradiction. Consequently, for all $Y \in 2^{(X)}$, we have

$$\sup_{z\in X}\inf_{x\in Y}\psi(x)(z) > r,$$

which proves that $c_3 \geq r$. From the arbitrariness of r this shows that $c_3 \geq \mu_c(X)$. Finally to show that $c_1 \geq c_3$ suppose that $c_3 > r$. Then there exists $\varphi_0 \in \prod_{x\in X} \mathcal{B}(x)$ such that, for all $Y \in 2^{(X)}$,

$$\sup_{z\in X}\inf_{x\in Y}\varphi_0(x)(z) > r.$$

Consequently, if, for all $Y \in 2^{(X)}$, we let

$$F_Y \doteqdot \left\{\inf_{x\in Y}\varphi_0(x) > r\right\},$$

then it follows that the collection $\left\{F_Y \mid Y \in 2^{(X)}\right\}$ is a basis for a filter $\mathcal{F}$ on X. Let $\mathcal{W}$ be an arbitrary ultrafilter finer than $\mathcal{F}$; then it follows that

$$\begin{aligned}
c_1 &= \sup_{\mathcal{U}\in\mathrm{U}(X)}\inf_{x\in X}\alpha\mathcal{U}(x)\\
&\geq \inf_{x\in X}\sup_{\varphi\in\mathcal{B}(x)}\sup_{W\in\mathcal{W}}\inf_{y\in W}\varphi(y)\\
&= \sup_{\varphi\in\prod_{x\in X}\mathcal{B}(x)}\inf_{x\in X}\sup_{W\in\mathcal{W}}\inf_{y\in W}\varphi(x)(y)\\
&\geq \inf_{x\in X}\sup_{W\in\mathcal{W}}\inf_{y\in W}\varphi_0(x)(y)\\
&\geq \inf_{x\in X}\inf_{y\in F_{\{x\}}}\varphi_0(x)(y)\\
&\geq r,
\end{aligned}$$

which by the arbitrariness of r shows that $c_1 \geq c_3$ and we are done.

□

Notice that the third expression for $\mu_c(X)$ constitutes a quantification of the characterization of compactness by means of neighborhoods, namely that if, for every $x \in X$, V_x is a neighborhood of x then there exists a finite set of points $x_1, \ldots, x_n \in X$ such that $X = V_{x_1} \cup \cdots \cup V_{x_n}$.

6.1.3 Theorem *If $(X, \delta_{\mathcal{T}})$ is a topological approach space, then the following are equivalent:*

1. *$(X, \mathcal{T})$ is compact,*
2. *$\mu_c(X) = 0$.*

Proof In view of the fact that the adherence operator in topological approach spaces attains only the values 0 and ∞, by definition of μ_c and upon invoking 2.1.3, the fact that $\mu_c(X) = 0$ precisely means that every filter has an adherence point.

□

6.1.4 Theorem *If (X, δ_d) is a p-metric approach space, then the following are equivalent:*

1. *(X, d) is totally bounded,*
2. *$\mu_c(X) = 0$.*

Proof It follows from 3.1.5 that

$$\mathcal{B}_d(x) \doteqdot \{d(x, \cdot)\} \qquad \forall x \in X$$

is a basis for the approach system of (X, δ_d). Applying 6.1.2 it therefore follows that

$$\mu_c(X) = \inf_{Y \in 2^{(X)}} \sup_{z \in X} \inf_{x \in Y} d(x, z).$$

Consequently the fact that $\mu_c(X) = 0$ means that for each $\epsilon > 0$ there exists a finite set $Y \in 2^{(X)}$ such that $X = \bigcup_{y \in Y} B(y, \epsilon)$, i.e. that X is totally bounded.

□

6.1.5 Theorem *If (X, δ_d) is a p-metric approach space, then the following are equivalent:*

1. *(X, d) is bounded,*
2. *$\mu_c(X) < \infty$.*

Proof This is perfectly analogous to 6.1.4.

□

6.1.6 Theorem *Suppose that (X, δ) is a uniform approach space and that δ is generated by the symmetric gauge basis $\mathcal{D}$. If $\mu_c(X) = 0$, then $(X, \mathcal{U}(\mathcal{D}))$ is totally bounded.*

Proof By 6.1.2, the fact that $\mu_c(X) = 0$ means that

$$0 = \sup_{\varphi \in \mathcal{D}^X} \inf_{Y \in 2^{(X)}} \sup_{z \in X} \inf_{x \in Y} \varphi(x)(x, z)$$

$$\geq \sup_{d \in \mathcal{D}} \inf_{Y \in 2^{(X)}} \sup_{z \in X} \inf_{x \in Y} d(x, z),$$

which in turn implies that, for any $d \in \mathcal{D}$, there exists a finite subset $Y \in 2^{(X)}$ such that

$$\bigcup_{x \in Y} B_d(x, \epsilon) = X.$$

Hence, all spaces (X, d) are totally bounded which means that $(X, \mathcal{U}(\mathcal{D}))$ is totally bounded.

□

The following example shows that the converse of 6.1.6 does not hold.

6.1.7 Example Let $X \doteqdot [-1, 1] \setminus \{0\}$ and let d_E stand for the Euclidean metric on X. If we consider the symmetric gauge basis $\mathcal{D} \doteqdot \{\alpha d_E \mid \alpha > 0\}$, then clearly the distance δ generated by $\mathcal{D}$ is topological ($\mathcal{T}_\delta$ is actually the Euclidean topology). Furthermore it is also clear that $\mathcal{U}(\mathcal{D})$ is the usual uniformity and, hence, $(X, \mathcal{U}(\mathcal{D}))$ is totally bounded. However, the restriction of the Euclidean neighborhood filter of 0 to X, say $\mathcal{F}$, does not converge in $(X, \mathcal{T}_\delta)$. Consequently, by 6.1.2 we have $\mu_c(X) = \infty$.

We now prove that, where both measures are defined, the Hausdorff measure of noncompactness and our measure of compactness μ_c coincide.

6.1.8 Theorem *If (X, δ_d) is a p-metric approach space, then $\mu_c(X) = m_H(X)$.*

Proof For any $\epsilon > 0$, we have $\mu_c(X) < \epsilon$ if and only if there exists a finite set $Y \in 2^{(X)}$ such that for all $z \in X$ there exists $x \in Y$ such that $z \in B(x, \epsilon)$, i.e. such that $X = \bigcup_{y \in Y} B(y, \epsilon)$. Thus $\mu_c(X) < \epsilon$ if and only if $m_H(X) < \epsilon$.

□

6.1.9 Theorem *If (X,δ) and (X',δ') are approach spaces and the function $f:(X,\delta)\longrightarrow(X',\delta')$ is a surjective contraction, then $\mu_c(X')\leq\mu_c(X)$.*

Proof Since every ultrafilter on X' is the image by f of an ultrafilter on X it follows from 1.9.2 that

$$\begin{aligned}\mu_c(X') &= \sup_{\mathcal{U}\in\mathbf{U}(X')}\inf_{y\in X'}\lambda'\mathcal{U}(y)\\ &\leq \sup_{\mathcal{U}\in\mathbf{U}(X)}\inf_{x\in X}\lambda' f(\mathcal{U})(f(x))\\ &\leq \sup_{\mathcal{U}\in\mathbf{U}(X)}\inf_{x\in X}\lambda\mathcal{U}(x)\\ &= \mu_c(X).\end{aligned}$$

□

6.1.10 Corollary *The continuous image of a compact topological space is compact.*

□

6.1.11 Corollary *The nonexpansive image of a totally bounded p-metric space is totally bounded.*

□

6.1.12 Corollary *The nonexpansive image of a bounded p-metric space is bounded.*

□

6.1.13 Theorem *If $(X_j,\delta_j)_{j\in J}$ is a family of approach spaces, then*

$$\mu_c\left(\prod_{j\in J}X_j\right)=\sup_{j\in J}\mu_c(X_j).$$

Proof It follows from 1.10.5 that if $\mathcal{U}$ is an ultrafilter on $\prod_{j\in J}X_j$, then $\alpha\mathcal{U}=\sup_{j\in J}\alpha_j(\mathrm{pr}_j(\mathcal{U}))\circ\mathrm{pr}_j$. Using this fact, the proof of the theorem now follows from the following calculation:

$$\begin{aligned}
\mu_c\left(\prod_{j\in J} X_j\right) &= \sup_{\mathcal{U}\in \mathbf{U}\left(\prod_{j\in J} X_j\right)} \inf_{x\in\prod_{j\in J} X_j} \alpha\mathcal{U}(x) \\
&= \sup_{\mathcal{U}\in \mathbf{U}\left(\prod_{j\in J} X_j\right)} \inf_{x\in\prod_{j\in J} X_j} \sup_{j\in J} \alpha_j(\mathrm{pr}_j(\mathcal{U}))(x_j) \\
&= \sup_{\mathcal{U}\in \mathbf{U}\left(\prod_{j\in J} X_j\right)} \sup_{j\in J} \inf_{z\in X_j} \alpha_j(\mathrm{pr}_j(\mathcal{U}))(z) \\
&= \sup_{j\in J} \sup_{\mathcal{U}\in \mathbf{U}(X_j)} \inf_{z\in X_j} \alpha_j\mathcal{U}(z) \\
&= \sup_{j\in J} \mu_c(X_j).
\end{aligned}$$

□

6.1.14 Corollary (Tychonoff theorem) *The product of a family of topological spaces is compact if and only if each factor space is compact.*

□

6.1.15 Corollary *The product of a finite family of p-metric spaces is totally bounded if and only if each factor space is totally bounded.*

□

We now give an example of the actual calculation of the measure of compactness of a subset of probability measures in the space $(\mathcal{P}, \delta_w)$ which we introduced in the foregoing chapter.

6.1.16 Example Let X be a separable metrizable space and let $(\mathcal{P}, \delta_w)$ be the space of probability measures on X equipped with the weak distance. In statistical applications one often encounters what, in the literature, are called contaminated sets of probability measures. Suppose that $\mathcal{H}$ is a weakly compact subset of $\mathcal{P}$ and that $\mathcal{K}$ is an arbitrary nonempty subset. Then we can consider the contaminated set

$$\mathcal{J} \doteqdot \{(1-\epsilon)P + \epsilon Q \mid P \in \mathcal{H}, Q \in \mathcal{K}\}.$$

Clearly, in general $\mathcal{J}$ need no longer be weakly compact. Nevertheless, from the point of view of statistics and if ϵ is sufficiently small, the set $\mathcal{J}$ may be indiscernible from the weakly compact set $\mathcal{H}$. Let us calculate $\mu_C(\mathcal{J})$. If $\mathbf{F}$ is a filter on $\mathcal{J}$ and $\mathcal{F} \in \mathbf{F}$, then we let

$$\mathcal{F}^* \doteqdot \{P \in \mathcal{H} \mid \exists Q \in \mathcal{K} : (1-\epsilon)P + \epsilon Q \in \mathcal{F}\}.$$

It is easily verified that $\{\mathcal{F}^* \mid \mathcal{F} \in \mathbf{F}\}$ generates a filter on $\mathcal{H}$ which we will denote by $\mathbf{F}^*$. Now let $P \in \mathcal{H}$ and $Q \in \mathcal{K}$; then it follows that if we let

$$R \doteqdot (1-\epsilon)P + \epsilon Q$$

we have that

$$\alpha_w \mathbf{F}(R) = \sup_{\mathcal{G} \in 2^{(\mathcal{T})}} \sup_{\mathcal{F} \in \mathbf{F}} \inf_{T \in \mathcal{F}} \sup_{G \in \mathcal{G}} (R(G) - T(G)) \vee 0.$$

Now fix $\mathcal{G} \in 2^{(\mathcal{T})}$, $\mathcal{F} \in \mathbf{F}$, and $T \in \mathcal{F}$ and let $U \in \mathcal{H}$ and $V \in \mathcal{K}$ be such that $T = (1-\epsilon)U + \epsilon V$; then

$$\begin{aligned}\sup_{G \in \mathcal{G}} (R(G) - T(G)) \vee 0 &= \sup_{G \in \mathcal{G}} ((1-\epsilon)(P(G) - U(G)) + \epsilon(Q(G) - V(G))) \vee 0 \\ &\leq \sup_{G \in \mathcal{G}} (P(G) - U(G)) \vee 0 + 2\epsilon\end{aligned}$$

and consequently,

$$\begin{aligned}\alpha_w \mathbf{F}(R) &\leq \sup_{\mathcal{G} \in 2^{(\mathcal{T})}} \sup_{\mathcal{F} \in \mathbf{F}} \inf_{U \in \mathcal{F}^*} \sup_{G \in \mathcal{G}} (P(G) - U(G)) \vee 0 \\ &= \alpha_w \mathbf{F}^*(P) + 2\epsilon.\end{aligned}$$

Now, since $\mathcal{H}$ is weakly compact, we find that

$$\begin{aligned}\mu_C(\mathcal{J}) &= \sup_{\mathbf{F} \in \mathbf{F}(\mathcal{J})} \inf_{R \in \mathcal{J}} \alpha_w \mathbf{F}(R) \\ &\leq \sup_{\mathbf{F} \in \mathbf{F}(\mathcal{J})} \inf_{P \in \mathcal{H}} \alpha_w \mathbf{F}^*(P) + 2\epsilon \\ &\leq \mu_C(\mathcal{H}) + 2\epsilon \\ &= 2\epsilon.\end{aligned}$$

□

6.2 CONNECTEDNESS

Our approach to connectedness follows the general theory of connectedness and disconnectedness in topological categories, as developed by Preuss [100]. For more information on connectedness concepts we also refer to the work of Arhangel'skii and Wiegandt in [3] and of Mrowka and Pervin in [89].

In order to formulate our definition of a measure of connectedness we need some preliminary concepts.

For each $\epsilon > 0$, we define the space D_ϵ to be the two-point set $\{0, \infty\}$ equipped with the ∞p-metric d_ϵ, where $d_\epsilon(0, \infty) \doteq \epsilon$. The space D_ϵ serves as a prototype for a disconnected space and the disconnectedness is quantified by the distance between the points 0 and ∞.

6.2.1 Definition An approach space (X, δ) is said to be *ϵ-connected* if the only contractions from (X, δ) to D_ϵ are constant functions.

Given an approach space (X, δ), we then define the *measure of connectedness* of X as

$$\mu_{cn}(X) \doteq \inf \{\epsilon > 0 \mid X \text{ is } \epsilon\text{-connected}\}.$$

6.2.2 Proposition *If (X, δ) is an approach space which is ϵ-connected and $\epsilon' > \epsilon$, then (X, δ) is ϵ'-connected.*

Proof This follows at once from the definition.

□

6.2.3 Example The space D_ϵ is ϵ'-connected, for all $\epsilon' > \epsilon$, but it is not ϵ-connected. Consequently $\mu_{cn}(D_\epsilon) = \epsilon$.

6.2.4 Theorem *If $(X, \delta_\mathcal{T})$ is a topological approach space, then the following are equivalent:*

1. $(X, \mathcal{T})$ *is connected,*

2. $\mu_{cn}(X) = 0$.

Proof To see that the first property implies the second one, suppose that for some $\epsilon > 0$ X is not ϵ-connected. Then there exists a contraction $f : X \longrightarrow D_\epsilon$ which is not constant. Let $X_0 \doteq f^{-1}(\{0\})$ and $X_\infty \doteq f^{-1}(\{\infty\})$. Then, for any

$x \in X_0$, it follows that $\epsilon \leq \delta_{\mathcal{T}}(x, X_\infty)$ which implies that $x \notin \mathrm{cl}_{\mathcal{T}}(X_\infty)$. Thus $X_0 \cap \mathrm{cl}_{\mathcal{T}}(X_\infty) = \emptyset$. Analogously $X_\infty \cap \mathrm{cl}_{\mathcal{T}}(X_0) = \emptyset$, which proves that both X_0 and X_∞ are open and thus $(X, \mathcal{T})$ is not connected.

To see that the second property implies the first one, let X be partitioned into two open sets X_0 and X_∞. Then it is easily verified that, for any $\epsilon > 0$, the map $f : X \longrightarrow D_\epsilon$ defined by $f(X_0) \doteqdot \{0\}$ and $f(X_\infty) \doteqdot \{\infty\}$ is a contraction. Thus $\mu_{cn}(X) = \infty$.

□

In order to prove our next result we recall that a p-metric space (X, d) is called *Cantor-connected*, Cantor [23], if, for any $\epsilon > 0$, any two points $x, y \in X$ can be "connected" by a so-called ϵ-chain, i.e. a finite number of points $x_0 = x, x_1, \ldots, x_n = y$ such that for all $i \in \{1, \ldots, n\}$: $d(x_{i-1}, x_i) \leq \epsilon$. It is well known that a space is Cantor-connected if and only if it cannot be partitioned into sets A and B such that $d(A, B) > 0$ (see e.g. Herrlich [49]).

6.2.5 Theorem *If (X, δ_d) is a p-metric approach space, then the following are equivalent:*

1. *(X, d) is Cantor-connected,*
2. *$\mu_{cn}(X) = 0$.*

Proof To see that the first property implies the second one suppose that for some $\epsilon > 0$ X is not ϵ-connected. Then there exists a contraction $f : X \longrightarrow D_\epsilon$ which is not constant. Let $X_0 \doteqdot f^{-1}(\{0\})$ and $X_\infty \doteqdot f^{-1}(\{\infty\})$. Then it follows that

$$\begin{aligned} d(X_0, X_\infty) &= \inf_{x \in X_0} \inf_{y \in X_\infty} d(x, y) \\ &\geq \inf_{x \in X_0} \inf_{y \in X_\infty} d_\epsilon(f(x), f(y)) \\ &= \inf_{x \in X_0} \inf_{y \in X_\infty} d_\epsilon(0, \infty) \\ &= \epsilon. \end{aligned}$$

To see that the second property implies the first one let X be partitioned into two sets X_0 and X_∞ such that $d(X_0, X_\infty) \geq \epsilon$. Then the map $f : X \longrightarrow D_\epsilon$, defined by $f(X_0) \doteqdot \{0\}$ and $f(X_\infty) \doteqdot \{\infty\}$, is a contraction and thus $\mu_{cn}(X) \geq \epsilon > 0$.

□

6.2.6 Examples We give two examples to illustrate the measure of connectedness in a topological and a metric setting.

1. If we consider the set of rational numbers $\mathbb{Q}$ to be equipped with the Euclidean topology, then this space is not connected; hence $\mu_{cn}(\mathbb{Q}) = \infty$. If on the other hand we consider it to be equipped with the Euclidean metric, then it is Cantor-connected and $\mu_{cn}(\mathbb{Q}) = 0$.

2. If we consider the natural numbers $\mathbb{N}$ to be equipped with the Euclidean topology, then this space is not connected; hence $\mu_{cn}(\mathbb{N}) = \infty$. If we consider it to be equipped with the Euclidean metric, then it is also not Cantor-connected but the measure of connectedness gives us $\mu_{cn}(\mathbb{N}) = 1$.

An alternative characterization for the measure of connectedness is given by the following result.

6.2.7 Proposition *For any approach space (X, δ) we have*

$$\mu_{cn}(X) = \sup_{A\in 2^X, A\neq\emptyset, A\neq X} \min\left\{\inf_{x\in A} \delta(x, X\setminus A), \inf_{x\in X\setminus A} \delta(x, A)\right\}.$$

Proof That X is not ϵ-connected is equivalent with the existence of a nonconstant contraction $f : X \longrightarrow D_\epsilon$. This in turn, however, is equivalent with the existence of a set A such that $A \neq \emptyset$, $A \neq X$, and such that, for all $x \in A$, $\delta(x, X \setminus A) \geq \epsilon$ and, for all $x \in X \setminus A$, $\delta(x, A) \geq \epsilon$.

□

We recall that a uniform space, with uniformity generated by a collection $\mathcal{D}$ of ∞p-metrics, is said to be *uniformly connected* (see Mrowka and Pervin [89], Arhangel'skii and Wiegandt [3]) if for all $A \subset X$, $d \in \mathcal{D}$, and $\epsilon > 0$ such that $A \neq \emptyset$ and $A \neq X$, we have

$$\{x \in X \mid \delta_d(x, X \setminus A) \leq \epsilon\} \bigcap \{x \in X \mid \delta_d(x, A) \leq \epsilon\} \neq \emptyset.$$

6.2.8 Theorem *Suppose that (X, δ) is a uniform approach space and that δ is generated by the symmetric gauge basis $\mathcal{D}$. If $\mu_{cn}(X) = 0$ then $(X, \mathcal{U}(\mathcal{D}))$ is uniformly connected.*

Proof Suppose that $(X, \mathcal{U}(\mathcal{D}))$ is not uniformly connected and let $A \subset X$, $d \in \mathcal{D}$, and $\epsilon > 0$ be such that $A \neq \emptyset$, $A \neq X$, and

$$\{x \in X \mid \delta_d(x, X \setminus A) \leq \epsilon\} \bigcap \{x \in X \mid \delta_d(x, A) \leq \epsilon\} = \emptyset.$$

Then it follows from 6.2.7 that

$$\begin{aligned}\mu_{cn}(X) &\geq \min\left\{\inf_{x\in A}\sup_{e\in\mathcal{D}}\delta_e(x, X\setminus A),\ \inf_{x\in X\setminus A}\sup_{e\in\mathcal{D}}\delta_e(x, A)\right\}\\ &\geq \min\left\{\inf_{x\in A}\delta_d(x, X\setminus A),\ \inf_{x\in X\setminus A}\delta_d(x, A)\right\}\\ &\geq \epsilon.\end{aligned}$$

□

The following example shows that the converse of 6.2.8 does not hold.

6.2.9 Example Let $X \doteqdot \mathbb{Q}$ and let d_E stand for the Euclidean metric on X. If we let $\mathcal{D} \doteqdot \{\alpha d_E \mid \alpha > 0\}$, then (as in 6.1.7) δ is the topological distance generated by the Euclidean topology and hence $\mu_{cn}(X) = \infty$. However, $\mathcal{U}(\mathcal{D})$ is the usual uniformity and, hence, $(X, \mathcal{U}(\mathcal{D}))$ is uniformly connected.

6.2.10 Proposition *If (X,δ) and (X',δ') are approach spaces, (X,δ) is ϵ-connected, and $f : (X,\delta) \longrightarrow (X',\delta')$ is a surjective contraction, then (X',δ') is ϵ-connected.*

Proof Suppose that X is ϵ-connected and that $h : X' \longrightarrow D_\epsilon$ is a contraction. Then $h \circ f : X \longrightarrow D_\epsilon$ is also a contraction which is therefore constant. Then it follows from the surjectivity of f that h too is constant. This proves that X' is ϵ-connected.

□

6.2.11 Theorem *If (X,δ) and (X',δ') are approach spaces and the function $f : (X,\delta) \longrightarrow (X',\delta')$ is a surjective contraction, then $\mu_{cn}(X') \leq \mu_{cn}(X)$.*

Proof This follows from the definition of μ_{cn} and 6.2.10.

□

6.2.12 Corollary *The continuous image of a connected topological space is connected.*

□

6.2.13 Corollary *The nonexpansive image of a Cantor-connected p-metric space is Cantor-connected.*

□

In order to give a product theorem for the measure of connectedness we need some more preliminary results, which are interesting in their own right.

6.2.14 Proposition *If $(X, \delta) \in |\mathbf{AP}|$, $Y \subset Z \subset \overline{Y} \subset X$, and Y is ϵ-connected, then Z is ϵ-connected.*

Proof Let $f : Z \to D_\epsilon$ be a contraction. Then $f_{|Y}$ is constant and thus there exists $p \in D_\epsilon$ such that $f(Y) = \{p\}$. Now if $z \in Z$, then $\delta(z, Y) = 0$ and, hence, $\delta_{d_\epsilon}(f(z), \{p\}) = 0$, i.e. $f(z) = p$. Thus, f is constant.

□

6.2.15 Definition If $\mathcal{A}$ is a collection of subsets of a set X, then $\mathcal{A}$ is said to be *chained* if, for any pair of sets $A, B \in \mathcal{A}$, there exists a finite collection $\{A_1, \ldots, A_n\} \subset \mathcal{A}$ such that $A_1 = A$, $A_n = B$, and for all $i \in \{1, \ldots, n-1\}$: $A_i \cap A_{i+1} \neq \emptyset$.

6.2.16 Proposition *Let (X, δ) be an approach space and let $\mathcal{A}$ be a chained collection of subsets of X. If each $A \in \mathcal{A}$ is ϵ_A-connected, then $\bigcup \mathcal{A}$ is $\left(\sup_{A \in \mathcal{A}} \epsilon_A\right)$-connected.*

Proof Suppose that $Y \doteqdot \bigcup \mathcal{A}$ is not $\left(\sup_{A \in \mathcal{A}} \epsilon_A\right)$-connected. Then there exists a subset $B \subset Y$ such that $\emptyset \neq B \neq Y$ and such that if we let $\epsilon \doteqdot \sup_{A \in \mathcal{A}} \epsilon_A$ then

$$\inf_{x \in B} \delta(x, Y \setminus B) \geq \epsilon \text{ and } \inf_{x \in Y \setminus B} \delta(x, B) \geq \epsilon.$$

Now suppose that there exists a set $A \in \mathcal{A}$ such that $A \not\subset B$ and $A \not\subset Y \setminus B$. Then it follows that

$$\inf_{x \in A \cap B} \delta(x, A \setminus B) \geq \inf_{x \in B} \delta(x, Y \setminus B) \geq \epsilon \geq \epsilon_A,$$

and

$$\inf_{x \in A \setminus B} \delta(x, A \cap B) \geq \inf_{x \in Y \setminus B} \delta(x, B) \geq \epsilon \geq \epsilon_A,$$

which is impossible. Consequently if we let $\mathcal{C} \doteqdot \{A \in \mathcal{A} \mid A \subset B\}$ and $\mathcal{D} \doteqdot \{A \in \mathcal{A} \mid A \subset Y \setminus B\}$, then $\mathcal{A} = \mathcal{C} \cup \mathcal{D}$. Since $\mathcal{A}$ is chained, this implies that either $\mathcal{C} = \emptyset$ or $\mathcal{D} = \emptyset$. Suppose, for example, that $\mathcal{D} = \emptyset$; then $\mathcal{A} = \mathcal{C}$ and it follows that $Y = B$ which is a contradiction.

□

6.2.17 Proposition *If $(X_j, \delta_j)_{j\in J}$ is a family of approach spaces such that for all $j \in J$, (X_j, δ_j) is ϵ_j-connected, then $\prod_{j\in J} (X_j, \delta_j)$ is $\left(\sup_{j\in J} \epsilon_j\right)$-connected.*

Proof Let $(z_j)_{j\in J}$ be a fixed point in the product space $\prod_{j\in J} X_j$. For any subset $K \subset J$, let

$$X_K \doteqdot \left\{(x_j)_{j\in J} \mid x_i = z_i \ \forall i \in J \setminus K\right\}.$$

By induction on the cardinality of K, making use of 6.2.16, one sees that, for any finite set K, X_K is $\left(\sup_{k\in K} \epsilon_k\right)$-connected. Now the collection

$$\mathcal{B} \doteqdot \{X_K \mid K \subset J,\ K \text{ finite}\}$$

is chained and, hence, it once again follows from 6.2.16 that $\bigcup \mathcal{B}$ is $\left(\sup_{j\in J} \epsilon_j\right)$-connected. Since $\bigcup \mathcal{B}$ is dense in $\prod_{j\in J} X_j$ it follows from 6.2.14 that $\prod_{j\in J} X_j$ is $\left(\sup_{j\in J} \epsilon_j\right)$-connected.

□

6.2.18 Theorem *If $(X_j, \delta_j)_{j\in J}$ is a family of approach spaces then*

$$\mu_{cn}\left(\prod_{j\in J} X_j\right) = \sup_{j\in J} \mu_{cn}(X_j).$$

Proof That $\mu_{cn}\left(\prod_{j\in J} X_j\right) \geq \sup_{j\in J} \mu_{cn}(X_j)$ follows from 6.2.11. The converse inequality follows from 6.2.17 since we have

$$\begin{aligned}\mu_{cn}\left(\prod_{j\in J} X_j\right) &= \inf\left\{\epsilon \mid \prod_{j\in J} X_j \ \ \epsilon\text{-connected}\right\}\\ &\leq \inf \bigcap_{j\in J} \{\epsilon \mid X_j \ \ \epsilon\text{-connected}\}\\ &= \sup_{j\in J} \mu_{cn}(X_j).\end{aligned}$$

□

6.2.19 Corollary *The product of a family of topological spaces is connected if and only if each factor space is connected.*

□

6.2.20 Corollary *The product of a finite family of p-metric spaces is Cantor-connected if and only if each factor space is Cantor-connected.*

□

6.2.21 Example If we consider the Cantor set $\mathbf{C}$ to be equipped with the Euclidean topology, then it is not connected; if we consider it to be equipped with the Euclidean metric, then it is also not Cantor-connected and it is easily seen that $\mu_{cn}(\mathbf{C}) = \frac{1}{3}$. If, however, we identify the Cantor set with the product $\{0,1\}^{\mathbb{N}}$ and equip $\{0,1\}$ with the Euclidean metric, then we can also equip $\mathbf{C}$ with the product distance. In that case, by 6.2.18, we have $\mu_{cn}(\mathbf{C}) = \sup_{n\in\mathbb{N}} \mu_{cn}(\{0,1\}) = 1$.

6.3 COMPLETENESS

Completeness is not a topological but a metric property and although it would be possible to quantify also metric properties in the setting of approach spaces we will not pursue that line of investigation in this work. Rather we will show that there exists a natural notion of completeness which, moreover, in the setting of uniform approach spaces, allows for a nice completion theory.

Intuitively, a Cauchy filter is a filter which has all the properties of a convergent filter, except that its limit points may be missing. This idea can be nicely captured using the concept of limit operator.

6.3.1 Definition A filter $\mathcal{F}$ in an approach space (X, δ) is called a *δ-Cauchy filter* or a *Cauchy filter* for short if

$$\inf_{x \in X} \lambda\mathcal{F}(x) = 0.$$

We say that $\mathcal{F}$ is *convergent* (to x) if

$$\lambda\mathcal{F}(x) = 0,$$

i.e. if it converges to x in the topological bicoreflection of (X, δ).

Hence in any approach space, a convergent filter is a Cauchy filter.

6.3.2 Proposition *If (X, δ) is a uniform approach space and $\mathcal{D}$ is a symmetric gauge basis which generates δ, then for a filter $\mathcal{F}$ on X the following are equivalent:*

1. *$\mathcal{F}$ is a Cauchy filter,*
2. $\forall \epsilon > 0\ \exists x \in X\ \forall d \in \mathcal{D}: \ B_d(x, \epsilon) \in \mathcal{F}$.

Proof From 1.8.31 it follows that $\mathcal{F}$ is a Cauchy filter if and only if

$$\inf_{x \in X} \sup_{d \in \mathcal{D}} \inf_{F \in \mathcal{F}} \sup_{y \in F} d(x, y) = 0,$$

from which the result follows immediately.

□

If (X, δ) is an arbitrary approach space, then we can consider its $p\mathbf{MET}^{\infty}$-bicoreflection (X, d_δ) and we have a metric notion of Cauchy filter at our disposal. If (X, δ) is a uniform approach space and δ is generated by the symmetric gauge basis $\mathcal{D}$, then we can consider the underlying uniform space $(X, \mathcal{U}(\mathcal{D}))$ and we have a uniform notion of Cauchy filter at our disposal. The following result gives the relationship between these concepts.

6.3.3 Proposition *1. If (X, δ) is an arbitrary approach space and $\mathcal{F}$ is a filter on X, then among the properties:*

(i) $\mathcal{F}$ is a d_δ-Cauchy filter,

(ii) $\mathcal{F}$ is a δ-Cauchy filter,

the implication (i) $\Rightarrow$ (ii) holds.

2. If (X, δ) is a uniform approach space, $\mathcal{D}$ is a symmetric gauge basis which generates δ, and $\mathcal{F}$ is a filter on X, then among the properties:

(i) $\mathcal{F}$ is a d_δ-Cauchy filter,

(ii) $\mathcal{F}$ is a δ-Cauchy filter,

(iii) $\mathcal{F}$ is a $\mathcal{U}(\mathcal{D})$-Cauchy filter,

the implications (i) $\Rightarrow$ (ii) $\Rightarrow$ (iii) hold.

Proof That $\mathcal{F}$ is a d_δ-Cauchy filter means that

$$\forall \epsilon > 0\, \exists x \in X : \bigcap_{d \in \mathcal{D}} B_d(x, \epsilon) \in \mathcal{F}.$$

That $\mathcal{F}$ is a $\mathcal{U}(\mathcal{D})$-Cauchy filter means that it is a d-Cauchy filter, for all $d \in \mathcal{D}$, i.e.

$$\forall \epsilon > 0,\ \forall d \in \mathcal{D}\, \exists x \in X : \ B_d(x, \epsilon) \in \mathcal{F}.$$

Hence, the result follows from 6.3.2.

□

6.3.4 Examples The implications in the foregoing proposition are strict in general.

1. Let $X \doteqdot \mathbb{R} \setminus \{0\}$ and let d_E stand for the Euclidean metric on X. If we let $\mathcal{D} \doteqdot \{\alpha d_E \mid \alpha > 0\}$, then we know (6.1.7) that the distance δ generated by $\mathcal{D}$ is topological. The restriction of the Euclidean neighborhood filter of 0 to X, say $\mathcal{F}$, does not converge in (X, δ) and hence, by 6.3.6, it is not a $\delta_{\mathcal{D}}$-Cauchy filter. However, $\mathcal{U}(\mathcal{D})$ is the usual uniformity on X and hence $\mathcal{F}$ is a $\mathcal{U}(\mathcal{D})$-Cauchy filter.

2. Let $\mathbb{R}$ be equipped with the Euclidean metric and consider the function space $\mathbb{R}^{\mathbb{R}}$ to be equipped with the product distance δ as in 5.2. For each $\alpha > 0$, consider the function

$$f_\alpha : \mathbb{R} \longrightarrow \mathbb{R} : x \longrightarrow \alpha x,$$

and let Ψ be the filter on $\mathbb{R}^{\mathbb{R}}$ generated by the basis

$$\{\{f_\beta \mid \beta \leq \alpha\} \mid \alpha > 0\}.$$

Since Ψ is convergent for the pointwise topology it follows that Ψ is a δ-Cauchy filter. However, since d_δ is the uniform ∞p-metric, Ψ is not a d_δ-Cauchy filter.

6.3.5 Proposition *If (X, δ) is a uniform approach space and $\mathcal{F}$ is a filter on X, then the following hold:*

1. *if $\mathcal{F}$ is a Cauchy filter, then $\lambda\mathcal{F} = \alpha\mathcal{F}$,*
2. *if $\mathcal{F}$ and $\mathcal{G}$ are Cauchy filters and $\mathcal{F} \subset \mathcal{G}$, then $\lambda\mathcal{F} = \lambda\mathcal{G}$.*

Proof To prove the first property let $\mathcal{D}$ be a symmetric gauge basis which generates δ and let $\mathcal{F}$ be a Cauchy filter. Then it follows from 6.3.3 that $\mathcal{F}$ is a $\mathcal{U}(\mathcal{D})$-Cauchy filter and consequently the result follows from 4.2.5.

The second property follows from the first one since, applying 1.8.35, we now have

$$\alpha\mathcal{F} = \alpha\mathcal{G} = \lambda\mathcal{G} = \lambda\mathcal{F}.$$

□

6.3.6 Proposition *Suppose $\mathcal{F} \in \mathbf{F}(X)$. Then the following hold:*

1. *if $(X, \delta_{\mathcal{T}})$ is a topological approach space, then $\mathcal{F}$ is a Cauchy filter if and only if it is convergent (in $(X, \mathcal{T})$),*
2. *if (X, δ_d) is an ∞p-metric approach space, then $\mathcal{F}$ is a Cauchy filter if and only if it is a Cauchy filter in (X, d).*

Proof The first property follows from the fact that in topological approach spaces the limit operator attains only the values 0 and ∞. The second property follows from 6.3.2.

□

6.3.7 Definition An approach space (X, δ) is said to be *complete* if every Cauchy filter in (X, δ) converges.

The following result emphasizes the fact that completeness is a metric and not a topological property.

6.3.8 Theorem *If $(X, \delta_{\mathcal{T}})$ is a topological approach space, then it is complete.*

Proof This follows from 6.3.2.

□

6.3.9 Theorem *If (X, δ_d) is an ∞p-metric approach space, then the following are equivalent:*

1. *(X, δ_d) is complete (as an approach space),*
2. *(X, d) is complete (as a metric space).*

Proof This follows from 3.2.1 and 6.3.2.

□

6.3.10 Theorem *Suppose that (X, δ) is a uniform approach space and that δ is generated by the symmetric gauge basis $\mathcal{D}$. If $(X, \mathcal{U}(\mathcal{D}))$ is complete, then (X, δ) is complete.*

Proof Let $\mathcal{F}$ be a Cauchy filter. It follows from 6.3.3 that $\mathcal{F}$ is a $\mathcal{U}(\mathcal{D})$-Cauchy filter. Since $(X, \mathcal{U}(\mathcal{D}))$ is complete it follows that $\mathcal{F}$ converges.

□

6.3.11 Theorem *Suppose that (X, δ) is a uniform approach space. The following properties are equivalent:*

1. *(X, δ) is complete,*
2. *(X, d_δ) is complete.*

Proof Suppose that (X, δ) is complete and that δ is generated by the symmetric gauge basis $\mathcal{D}$. Let $(x_n)_n$ be a Cauchy sequence in (X, d_δ). Then we have that

$$\forall \epsilon > 0 \; \exists n_0 \; \forall p, q \geq n_0 \; \forall d \in \mathcal{D} : \; d(x_p, x_q) \leq \epsilon.$$

This implies that

$$\lambda \langle (x_n)_n \rangle (x_{n_0}) = \sup_{d \in \mathcal{D}} \inf_m \sup_{k \geq m} d(x_{n_0}, x_k) \leq \epsilon,$$

and hence it follows that $\langle (x_n)_n \rangle$ is a δ-Cauchy filter. Since (X, δ) is complete this implies that there exists $x \in X$ such that $\lambda \langle (x_n)_n \rangle (x) = 0$, i.e.

$$\forall d \in \mathcal{D} : \; \lim_{n \to \infty} d(x_n, x) = 0.$$

Now let $\epsilon > 0$; then from the fact that $(x_n)_n$ is a d_δ-Cauchy sequence it follows that we can find n_0 such that

$$\forall d \in \mathcal{D}, \forall p, q \geq n_0 : \; d(x_p, x_q) \leq \epsilon.$$

Taking the limit for $q \to \infty$ it follows that

$$\forall d \in \mathcal{D}, \forall p \geq n_0 : \; d(x_p, x) \leq \epsilon,$$

which implies that $(x_n)_n \to x$ in (X, d_δ).

Conversely, suppose that (X, d_δ) is complete and let $\mathcal{F}$ be a δ-Cauchy filter. Then

$$\inf_{x \in X} \lambda\mathcal{F}(x) = \inf_{x \in X} \sup_{d \in \mathcal{D}} \inf_{F \in \mathcal{F}} \sup_{y \in F} d(x, y) = 0,$$

which implies that

$$\forall n > 0 \; \exists x_n \in X \; \forall d \in \mathcal{D} : \; B_d\left(x_n, \frac{1}{n}\right) \in \mathcal{F}.$$

Since the intersection of any two of these balls is nonempty this implies that for any $p, q > 0$

$$d_\delta(x_p, x_q) = \sup_{d \in \mathcal{D}} d(x_p, x_q) \leq \frac{1}{p} + \frac{1}{q},$$

from which it follows that $(x_n)_n$ is a d_δ-Cauchy sequence. By the completeness of (X, d_δ) it follows that there exists $x \in X$ such that $(x_n)_n \stackrel{d_\delta}{\to} x$. It then follows that

$$\begin{aligned}
\lambda\mathcal{F}(x) &= \sup_{d \in \mathcal{D}} \inf_{F \in \mathcal{F}} \sup_{y \in F} d(x, y) \\
&\leq \sup_{d \in \mathcal{D}} \inf_{n > 0} \sup_{y \in B_d\left(x_n, \frac{1}{n}\right)} d(x, y) \\
&\leq \sup_{d \in \mathcal{D}} \inf_{n > 0} \left(d(x, x_n) + \frac{1}{n} \right) \\
&= 0,
\end{aligned}$$

and hence $\mathcal{F}$ converges in (X, δ).

□

In 6.1.6 and 6.3.10 we have seen that as far as completeness and total boundedness are concerned, when these properties are considered independently only one-sided implications are possible for the properties with respect to the various structures involved. When combined, however, we obtain the following result.

6.3.12 Theorem *Suppose that (X, δ) is a uniform approach space. The following properties are equivalent:*

1. $(X, \mathcal{T}_\delta)$ *is compact,*
2. (X, δ) *is complete and* $\mu_c(X) = 0$.

Proof Let δ be generated by the symmetric gauge basis $\mathcal{D}$. If $(X, \mathcal{T}_\delta)$ is compact, then $(X, \mathcal{U}(\mathcal{D}))$ is complete and it follows from 6.3.10 that (X, δ) is complete. Furthermore it then follows from 6.1.9 that $\mu_c(X) = 0$.

Conversely, if $\mathcal{U}$ is an ultrafilter on X it follows from the fact that $\mu_c(X) = 0$ and the definition of μ_c that $\mathcal{U}$ is a Cauchy filter. From the completeness of (X, δ) it then follows that $\mathcal{U}$ converges. Hence, $(X, \mathcal{T}_\delta)$ is compact.

□

We now look at some examples of complete approach spaces other than topological spaces and complete metric spaces.

6.3.13 Example First we look at an example involving the function spaces introduced in the foregoing chapter. Incidentally this example also shows that the result of 6.3.10 is best possible. In particular it provides us with a complete uniform approach space for which the associated uniform space is not complete.

Let a set X and a metric space (Y, d) be given. The following is an immediate consequence of 6.3.11.

If $\mathcal{Z} \subset Y^X$ *is equipped with the distance* δ *of pointwise convergence then among the properties*

1. $(\mathcal{Z}, \delta)$ *is complete,*
2. $(\mathcal{Z}, d_\delta)$ *is complete,*
3. $(\mathcal{Z}, \mathcal{U}(\delta))$ *is complete,*

we have the following relationships: $1 \Leftrightarrow 2 \Leftarrow 3$.

Further to the conditions above if X is a topological space let $\mathcal{C}(X, Y)$ stand for the set of all continuous functions from X to Y, if X is a uniform space let $\mathcal{C}_u(X, Y)$ stand for the set of uniformly continuous maps from X to Y, and if X is a metric space let $\mathcal{H}(X, Y)$ stand for the set of nonexpansive maps from X to Y. In all cases we equip the function space with the pointwise distance. Since all these spaces are complete for the metric of uniform convergence, the following is a consequence of the equivalence of 1 and 2 above.

If X is a topological (respectively uniform or metric) space, then $\mathcal{C}(X,Y)$ (respectively $\mathcal{C}_u(X,Y)$ or $\mathcal{H}(X,Y)$), equipped with the distance of pointwise convergence, is complete.

It is well known that for the uniformity of pointwise convergence none of these subspaces of Y^X is complete. In spite of the fact that the **TOP**-bicoreflection of the pointwise distance, in all cases, is the pointwise topology and the associated uniformity is the pointwise uniformity, this does not contradict the fact that these subspaces are complete for the pointwise distance. The explanation lies in the fact that the notion of δ-Cauchy filter precisely eliminates the "bad" pointwise Cauchy filters from consideration.

6.3.14 Example Now let us look at the example concerning hyperspaces, where we obtain a similar result. In Beer [10] it was shown that if (X,d) is complete and separable, then $(CL(X),\mathcal{T}_{W_d})$ is Polish, i.e. it is separable and completely metrizable. This means that among all metrics which metrize $(CL(X),\mathcal{T}_{W_d})$ a complete one can be found. However, from 6.3.11 we obtain the following result.

For any metric space (X,d) the following are equivalent:

1. *$(CL(X),\delta_{W_d})$ is complete,*
2. *$(CL(X),h_d)$ is complete,*
3. *(X,d) is complete.*

The equivalence of 2 and 3 can be found, for example, in Kuratowski [63]. This result shows that the natural distance δ_{W_d}, which "distancizes" the Wijsman topology, without further assumptions such as separability, is itself already complete in the case when (X,d) is complete.

7
COMPLETION

The fact that **AP** contains both **TOP** and p**MET**$^\infty$ as full and isomorphism-closed subconstructs implies that we have to consider both the notions of completion and of compactification, at least for uniform approach spaces. In the present chapter we construct a completion which coincides with the usual completion in the case of p-metric spaces and which is an epireflection from the subconstruct of Hausdorff uniform approach spaces to the subconstruct of complete Hausdorff uniform approach spaces. In the second section we look at the relationship between the p**MET**$^\infty$-coreflection of this completion and the usual metric completion in the special case of function spaces.

7.1 CONSTRUCTION

In order to construct a completion for uniform approach spaces we proceed as in the case of the metric or the uniform completion, by first adding new points to the space and then extending the structure.

Let (X, δ) be a uniform approach space. For any δ-Cauchy filter $\mathcal{F}$, put

$$C(\mathcal{F}) \doteqdot \{\mathcal{G} \in \mathbf{F}(X) \mid \mathcal{G} \ \delta\text{-Cauchy}, \ \mathcal{G} \subset \mathcal{F}\},$$

and let

$$\mathcal{M}_{\mathcal{F}} \doteqdot \bigcap_{\mathcal{G} \in C(\mathcal{F})} \mathcal{G}.$$

7.1.1 Proposition *Let (X, δ) be a uniform approach space and let $\mathcal{F}$ be a δ-Cauchy filter. Then $\mathcal{M}_{\mathcal{F}}$ is the smallest δ-Cauchy filter coarser than $\mathcal{F}$ and $\lambda\mathcal{M}_{\mathcal{F}} = \lambda\mathcal{F}$.*

Proof It follows from (L3) and 6.3.5 that

$$\lambda\mathcal{M}_{\mathcal{F}} = \sup_{\mathcal{G} \in C(\mathcal{F})} \lambda\mathcal{G} = \sup_{\mathcal{G} \in C(\mathcal{F})} \lambda\mathcal{F} = \lambda\mathcal{F}.$$

By definition this then also implies that $\mathcal{M}_{\mathcal{F}}$ is a δ-Cauchy filter, which by construction is of course the smallest one coarser than $\mathcal{F}$.

□

Given δ-Cauchy filters $\mathcal{F}$ and $\mathcal{G}$ we will say that they are *equivalent* if $\mathcal{F} \cap \mathcal{G}$ is a δ-Cauchy filter, and we denote this by $\mathcal{F} \sim \mathcal{G}$. This is indeed an equivalence relation. Symmetry and reflexivity are evident. If $\mathcal{F} \sim \mathcal{G}$ and $\mathcal{G} \sim \mathcal{H}$, then it follows from 6.3.5 that $\lambda\mathcal{F} = \lambda(\mathcal{F} \cap \mathcal{G}) = \lambda\mathcal{G} = \lambda(\mathcal{G} \cap \mathcal{H}) = \lambda\mathcal{H}$. Hence, $\lambda(\mathcal{F} \cap \mathcal{H}) = \lambda\mathcal{F} \vee \lambda\mathcal{H} = \lambda\mathcal{F}$ which implies that $\mathcal{F} \sim \mathcal{H}$.

7.1.2 Proposition *Let (X, δ) be a uniform approach space. Then the δ-Cauchy filters $\mathcal{F}$ and $\mathcal{G}$ are equivalent if and only if $\lambda\mathcal{F} = \lambda\mathcal{G}$.*

Proof If $\mathcal{F} \sim \mathcal{G}$, then it follows from 6.3.5 that $\lambda\mathcal{F} = \lambda(\mathcal{F} \cap \mathcal{G}) = \lambda\mathcal{G}$. Conversely if $\lambda\mathcal{F} = \lambda\mathcal{G}$, then $\lambda(\mathcal{F} \cap \mathcal{G}) = \lambda\mathcal{F} \vee \lambda\mathcal{G} = \lambda\mathcal{F}$ which implies that $\mathcal{F} \cap \mathcal{G}$ is a δ-Cauchy filter and hence that $\mathcal{F} \sim \mathcal{G}$.

□

7.1.3 Proposition *Let (X, δ) be a uniform approach space. Then, for any δ-Cauchy filter $\mathcal{F}$, $\mathcal{M}_{\mathcal{F}}$ is the minimal element of the set of all δ-Cauchy filters which are equivalent with $\mathcal{F}$.*

Proof That $\mathcal{M}_{\mathcal{F}}$ is equivalent with $\mathcal{F}$ follows from 7.1.1 and 7.1.2. If $\mathcal{G}$ is equivalent with $\mathcal{F}$ then by 7.1.1, $\mathcal{M}_{\mathcal{F}} \subset \mathcal{F} \cap \mathcal{G} \subset \mathcal{G}$ and, hence, $\mathcal{M}_{\mathcal{F}}$ is minimal.

□

A filter of type $\mathcal{M}_{\mathcal{F}}$ will be called a *minimal δ-Cauchy filter* or *minimal Cauchy filter* if no confusion can occur.

7.1.4 Proposition *If (X, δ) is a uniform approach space, then, for any $x \in X$, the neighborhood filter $\mathcal{N}(x)$ in the* **TOP**-*bicoreflection is a minimal δ-Cauchy filter.*

Proof Since $\lambda\mathcal{N}(x)(x) = 0$ it follows that $\mathcal{N}(x)$ is a δ-Cauchy filter. If $\mathcal{G} \subset \mathcal{N}(x)$ is a δ-Cauchy filter, then it follows from 6.3.5 that $\lambda\mathcal{G}(x) = \lambda\mathcal{N}(x)(x) = 0$ and hence $\mathcal{G}$ converges to x in the **TOP**-bicoreflection of (X, δ). Consequently, $\mathcal{G} \supset \mathcal{N}(x)$ which proves that $\mathcal{N}(x)$ is minimal.

□

Now if δ is generated by the symmetric gauge basis $\mathcal{D}$ and $\mathcal{F}$ is a δ-Cauchy filter, then it follows from 6.3.3 that it is also a $\mathcal{U}(\mathcal{D})$-Cauchy filter. Consequently, there is a minimal $\mathcal{U}(\mathcal{D})$-Cauchy filter contained in $\mathcal{F}$. This filter, which we denote by $\mathcal{F}^{\mathcal{D}}$, is generated by the basis

$$\{F_d^\epsilon \mid d \in \mathcal{D}, F \in \mathcal{F}, \epsilon > 0\},$$

where, for any $d \in \mathcal{D}, F \in \mathcal{F}$, and $\epsilon > 0$,

$$F_d^\epsilon \doteqdot \{x \in X \mid \delta_d(x, F) \leq \epsilon\}.$$

7.1.5 Proposition *If (X, δ) is a uniform approach space, $\mathcal{D}$ is a symmetric gauge basis which generates δ, and $\mathcal{F}$ is a δ-Cauchy filter, then $\mathcal{F}^{\mathcal{D}} = \mathcal{M}_{\mathcal{F}}$.*

Proof Let $\mathcal{F}$ be a δ-Cauchy filter. Then, for any $x \in X$, we have

$$\begin{aligned}\lambda\mathcal{F}^{\mathcal{D}}(x) &= \sup_{d\in\mathcal{D}} \inf_{d'\in\mathcal{D}} \inf_{\epsilon>0} \inf_{F\in\mathcal{F}} \sup_{z\in F_{d'}^\epsilon} d(x, z) \\ &\leq \sup_{d\in\mathcal{D}} \inf_{\epsilon>0} \inf_{F\in\mathcal{F}} \sup_{z\in F_d^\epsilon} d(x, z) \\ &\leq \sup_{d\in\mathcal{D}} \inf_{F\in\mathcal{F}} \inf_{\epsilon>0} \left(\sup_{z\in F} d(x, z) + \epsilon\right) \\ &= \lambda\mathcal{F}(x).\end{aligned}$$

Consequently, it follows that $\mathcal{F}^{\mathcal{D}}$ too is a δ-Cauchy filter. Since $\mathcal{F}^{\mathcal{D}} \subset \mathcal{F}$, the minimality of $\mathcal{M}_{\mathcal{F}}$ implies that $\mathcal{M}_{\mathcal{F}} \subset \mathcal{F}^{\mathcal{D}}$. Conversely, $\mathcal{M}_{\mathcal{F}}$ is also a $\mathcal{U}(\mathcal{D})$-Cauchy filter, and thus it follows from the minimality of $\mathcal{F}^{\mathcal{D}}$ that $\mathcal{F}^{\mathcal{D}} \subset \mathcal{M}_{\mathcal{F}}$.

□

Given (X, δ) we denote by $\hat{X}$ the set of all minimal δ-Cauchy filters. Note that it follows from 7.1.5 that a minimal δ-Cauchy filter has an open base in the **TOP**-bicoreflection.

7.1.6 Corollary *If (X, δ) is a Hausdorff uniform approach space, then the map*

$$e_X : X \longrightarrow \hat{X} : x \longrightarrow \mathcal{N}(x)$$

is an injection and thus the set X can be identified with the subset $e_X(X)$ of $\hat{X}$.

□

If δ is generated by the symmetric gauge basis $\mathcal{D}$ then for any $d \in \mathcal{D}$ we define $\hat{d}$ by

$$\hat{d}(\mathcal{F}, \mathcal{G}) = \sup_{F \in \mathcal{F}} \sup_{G \in \mathcal{G}} \inf_{x \in F} \inf_{y \in G} d(x, y) \qquad \mathcal{F}, \mathcal{G} \in \hat{X}.$$

It follows from 7.1.5 that minimal δ-Cauchy filters are also minimal $\mathcal{U}(\mathcal{D})$-Cauchy filters and hence the function $\hat{d}$ is well defined.

7.1.7 Proposition *If (X, δ) is a uniform approach space and $\mathcal{D}$ is a symmetric gauge basis which generates δ, then the set*

$$\hat{\mathcal{D}} \doteqdot \left\{ \hat{d} \mid d \in \mathcal{D} \right\}$$

is a symmetric gauge basis on $\hat{X}$.

Proof It suffices to prove that $\hat{\mathcal{D}}$ is upwards directed. This follows from the fact that, for any $d, e \in \mathcal{D}$, we have $\hat{d} \vee \hat{e} \leq \widehat{d \vee e}$, as is easily verified.

□

The distance generated by $\hat{\mathcal{D}}$ will be denoted by $\hat{\delta}$ and analogously for the associated structures.

7.1.8 Proposition *If (X, δ) is a Hausdorff uniform approach space, then the map*

$$e_X : (X, \delta) \longrightarrow \left(\hat{X}, \hat{\delta}\right) : x \longrightarrow \mathcal{N}(x)$$

is an embedding and thus (X, δ) can be identified with the subspace $\left(e_X(X), \hat{\delta}_{|X}\right)$ of $\left(\hat{X}, \hat{\delta}\right)$.

Proof Together with 7.1.6 it suffices to notice that if $\mathcal{D}$ is a symmetric gauge basis which generates δ, then, for any $d \in \mathcal{D}$, we have

$$\begin{aligned}\hat{d}(\mathcal{N}(x), \mathcal{N}(y)) &= \sup_{F \in \mathcal{N}(x)} \sup_{G \in \mathcal{N}(y)} \inf_{u \in F} \inf_{v \in G} d(u, v) \\ &= d(x, y).\end{aligned}$$

□

7.1.9 Corollary *If (X,δ) is a complete, Hausdorff uniform approach space, then $\left(\hat{X},\hat{\delta}\right)$ is isomorphic to (X,δ) via the bijection*

$$e_X : (X,\delta) \longrightarrow \left(\hat{X},\hat{\delta}\right) : x \longrightarrow \mathcal{N}(x).$$

□

7.1.10 Proposition *If (X,δ) is a uniform approach space, $\mathcal{C} \in \hat{X}$, and $x \in X$, then*

$$\lambda\mathcal{C}(x) = \hat{\delta}(\mathcal{C},\{\mathcal{N}(x)\}).$$

Proof Let $\mathcal{D}$ be a symmetric gauge basis which generates δ. Applying 6.3.5 this follows from

$$\begin{aligned}\hat{\delta}(\mathcal{C},\{\mathcal{N}(x)\}) &= \sup_{d\in\mathcal{D}}\sup_{d'\in\mathcal{D}}\sup_{\epsilon>0}\sup_{A\in\mathcal{C}}\inf_{z\in B_{d'}(x,\epsilon)}\inf_{y\in A} d(z,y)\\ &= \sup_{d\in\mathcal{D}}\sup_{\epsilon>0}\sup_{A\in\mathcal{C}}\inf_{z\in B_{d}(x,\epsilon)}\inf_{y\in A} d(z,y)\\ &= \sup_{d\in\mathcal{D}}\sup_{A\in\mathcal{C}}\inf_{y\in A} d(x,y)\\ &= \alpha\mathcal{C}(x)\\ &= \lambda\mathcal{C}(x).\end{aligned}$$

□

7.1.11 Proposition *The subset $e_X(X)$ is dense in $\left(\hat{X},\mathcal{T}_{d_{\hat{\delta}}}\right)$, i.e. with respect to the topology generated by the* p**MET**$^\infty$*-bicoreflection of $\hat{\delta}$.*

Proof By 7.1.10 we have, for any $x \in X$ and $\mathcal{C} \in \hat{X}$,

$$\begin{aligned}d_{\hat{\delta}}(\mathcal{C},\mathcal{N}(x)) &= \hat{\delta}(\mathcal{C},\{\mathcal{N}(x)\})\\ &= \lambda\mathcal{C}(x),\end{aligned}$$

and thus, since $\mathcal{C}$ is δ-Cauchy,

$$\begin{aligned}d_{\hat{\delta}}(\mathcal{C},e_X(X)) &= \inf_{x\in X} d_{\hat{\delta}}(\mathcal{C},\mathcal{N}(x))\\ &= \inf_{x\in X}\lambda\mathcal{C}(x)\\ &= 0.\end{aligned}$$

□

For any filter $\mathcal{F}$ on X, we will denote the filter generated on $\hat{X}$ by the basis $e_X(\mathcal{F})$ by $\hat{\mathcal{F}}$. Notice that if $\mathcal{F}$ is a δ-Cauchy filter, then $\hat{\mathcal{F}}$ is a $\hat{\delta}$-Cauchy filter.

7.1.12 Proposition *If (X, δ) is a uniform approach space and $\mathcal{C} \in \hat{X}$, then*

$$\hat{\lambda}\hat{\mathcal{C}}(\mathcal{C}) = 0.$$

Proof Let $\mathcal{D}$ be a symmetric gauge basis which generates δ. Fix $\epsilon > 0$; then, since $\mathcal{C}$ is δ-Cauchy, it follows from 6.3.2 that we can find $x \in X$ such that, for all $d' \in \mathcal{D}$, we have $B_{d'}(x, \epsilon) \in \mathcal{C}$. Consequently, we obtain

$$\begin{aligned}
\hat{\lambda}\hat{\mathcal{C}}(\mathcal{C}) &= \sup_{d\in\mathcal{D}} \inf_{A\in\mathcal{C}} \sup_{a\in A} \hat{d}(\mathcal{N}(a), \mathcal{C}) \\
&= \sup_{d\in\mathcal{D}} \inf_{A\in\mathcal{C}} \sup_{a\in A} \sup_{A'\in\mathcal{C}} \inf_{b\in A'\cap A} d(a, b) \\
&\leq \sup_{d\in\mathcal{D}} \inf_{d'\in\mathcal{D}} \sup_{a\in B_{d'}(x,\epsilon)} \sup_{A'\in\mathcal{C}} \inf_{b\in A'\cap B_{d'}(x,\epsilon)} d(a, b) \\
&\leq \sup_{d\in\mathcal{D}} \sup_{a\in B_d(x,\epsilon)} \sup_{A'\in\mathcal{C}} \inf_{b\in A'\cap B_d(x,\epsilon)} d(a, b) \\
&\leq 2\epsilon
\end{aligned}$$

which by the arbitrariness of ϵ proves our claim.

□

In what follows we will denote the subconstruct of **UAP** consisting of all Hausdorff spaces by $\mathbf{UAP}_2$ and the subconstruct of $\mathbf{UAP}_2$ consisting of all complete spaces by $c\mathbf{UAP}_2$.

In order to prove our main theorem of this section we first have to characterize the epimorphisms in $\mathbf{UAP}_2$.

7.1.13 Proposition *A function is an epimorphism in $\mathbf{UAP}_2$ if and only if it is a contraction which has a dense image.*

Proof The if part follows from the analogous result in **TOP**. To show the only-if part suppose that (X, δ_X) and (Y, δ_Y) are uniform approach spaces, where δ_X is generated by $\mathcal{D}_X$ and δ_Y is generated by the collection of bounded p-metrics $\mathcal{D}_Y$. Further let

$$f : (X, \delta_X) \longrightarrow (Y, \delta_Y)$$

be a nondense contraction. Then there exists $y_0 \in Y$ such that

$$\delta_Y(y_0, f(X)) > 0$$

and hence there further exists $d_0 \in \mathcal{D}_Y$ such that

$$\inf_{x \in X} d_0(y_0, f(x)) > 0.$$

Now consider $[0, \infty[$ equipped with the Euclidean metric d_E and define the map

$$\theta : (Y, \delta_Y) \longrightarrow ([0, \infty[, \delta_{d_E}) : y \longrightarrow \inf_{x \in X} d_0(y_0, f(x)).$$

Since, for all $z \in Y, Z \in 2^Y$, we have

$$\begin{aligned}
\delta_{d_E}(\theta(z), \theta(Z)) &= \inf_{z' \in Z} \left| \inf_{x \in X} d_0(z, f(x)) - \inf_{x \in X} d_0(z', f(x)) \right| \\
&\leq \inf_{z' \in Z} \sup_{x \in X} \left| d_0(z, f(x)) - d_0(z', f(x)) \right| \\
&\leq \inf_{z' \in Z} d_0(z, z') \\
&\leq \delta_Y(z, Z),
\end{aligned}$$

it follows that θ is a contraction. Clearly $\theta \circ f = 0$ whereas $\theta(y_0) \neq 0$. The constant map $\eta = 0$ is also a contraction and, since $\theta \neq \eta$ whereas $\theta \circ f = \eta \circ f$, it follows that f is not an epimorphism.

□

7.1.14 Proposition *A function is an extremal monomorphism in* $\mathbf{UAP}_2$ *if and only if it is an injective contraction which has a closed image.*

Proof This goes in exactly the same way as the proof for $\mathbf{TOP}_2$ (the subconstruct of $\mathbf{TOP}$ with objects all Hausdorff spaces) and we therefore leave this to the reader.

□

7.1.15 Theorem $\mathrm{c}\mathbf{UAP}_2$ *is an epireflective subconstruct of* $\mathbf{UAP}_2$. *For any* $(X, \delta) \in |\mathbf{UAP}_2|$,

$$e_X : (X, \delta) \longrightarrow \left(\hat{X}, \hat{\delta}\right) : x \longrightarrow \mathcal{N}(x)$$

is an epireflection of (X, δ) *in* $\mathrm{c}\mathbf{UAP}_2$.

Proof 1. $\left(\hat{X}, \hat{\delta}\right)$ *is Hausdorff.* This is clear since, by construction, its $\mathbf{TOP}$-bicoreflection is a subspace of the topological space underlying the uniform completion of $(X, \mathcal{U}(\mathcal{D}))$.

2. $\left(\hat{X},\hat{\delta}\right)$ *is complete.* To prove this, first of all notice that we can restrict our attention to showing that minimal $\hat{\delta}$-Cauchy filters are convergent. Let $\mathbf{\Phi}$ be such a minimal $\hat{\delta}$-Cauchy filter. Then, since it has an open base, it has a trace on X, say $\mathcal{F}$. Let $\epsilon > 0$. Then it follows from 7.1.12 that we can find $\mathcal{C} \in \hat{X}$ such that

$$\hat{\lambda}\mathbf{\Phi}(\mathcal{C}) \leq \epsilon$$

and next we can find $x \in X$ such that

$$\lambda\mathcal{C}(x) \leq \epsilon.$$

Then from the fact that $\mathbf{\Phi} \subset \hat{\mathcal{F}}$, that $\lambda\mathcal{F}(x) = \hat{\lambda}\hat{\mathcal{F}}(\mathcal{N}(x))$, and upon applying 4.2.1 and 7.1.10 it follows that

$$\begin{aligned}\lambda\mathcal{F}(x) &\leq \hat{\lambda}\mathbf{\Phi}(\mathcal{N}(x))\\ &\leq \hat{\lambda}\mathbf{\Phi}(\mathcal{C}) + \hat{\delta}(\mathcal{C}, \{\mathcal{N}(x)\})\\ &\leq 2\epsilon,\end{aligned}$$

which by the arbitrariness of ϵ implies that $\mathcal{F}$ is a δ-Cauchy filter.

Now, since $\mathbf{\Phi}$ and $\hat{\mathcal{F}}$ are $\hat{\delta}$-Cauchy filters and since $(\widehat{\mathcal{M}_{\mathcal{F}}}) \subset \hat{\mathcal{F}}$ and $\mathbf{\Phi} \subset \hat{\mathcal{F}}$, it follows from 6.3.5 and 7.1.12 that

$$\begin{aligned}\hat{\lambda}\mathbf{\Phi}(\mathcal{M}_{\mathcal{F}}) &= \hat{\lambda}\hat{\mathcal{F}}(\mathcal{M}_{\mathcal{F}})\\ &= \hat{\lambda}(\widehat{\mathcal{M}_{\mathcal{F}}})(\mathcal{M}_{\mathcal{F}})\\ &= 0.\end{aligned}$$

This proves that $\left(\hat{X},\hat{\delta}\right)$ is complete.

3. $e_X : (X,\delta) \longrightarrow \left(\hat{X},\hat{\delta}\right) : x \longrightarrow \mathcal{N}(x)$ *determines an epireflection.* That e_X is an epimorphism follows from 7.1.13 and the obvious fact that $e_X(X)$ is dense in $\left(\hat{X},\hat{\delta}\right)$.

Now suppose that $(Y,\delta_Y) \in |\mathrm{c}\mathbf{UAP_2}|$ and that

$$f : (X,\delta) \longrightarrow (Y,\delta_Y)$$

is a contraction. If $\mathcal{C} \in \hat{X}$, then $f(\mathcal{C})$ is a δ_Y-Cauchy filter and thus there exists a unique point $y_{\mathcal{C}} \in Y$ such that $\lambda_Y f(\mathcal{C})(y_{\mathcal{C}}) = 0$. We define

$$\hat{f} : \hat{X} \longrightarrow Y : \mathcal{C} \longrightarrow y_{\mathcal{C}}.$$

Since, for any $x \in X$, we have that $\lambda_Y f(\mathcal{N}(x))(f(x)) = 0$ it follows that the diagram

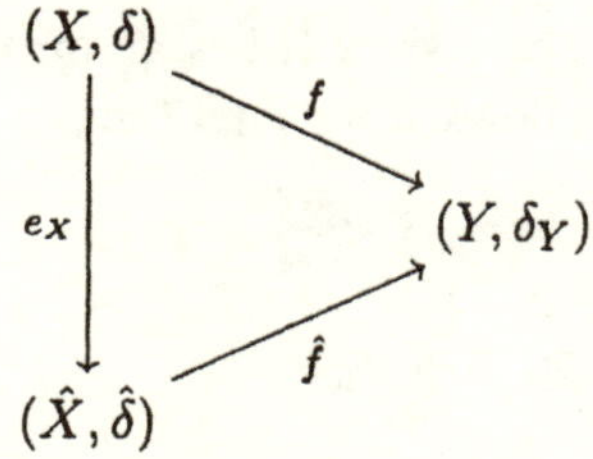

commutes. Moreover, since e_X is an epimorphism, $\hat{f}$ is the only function which makes the diagram commute. So all that remains to be shown is that it is a contraction. Let $\epsilon > 0$ and let us denote by

$$\overline{\epsilon} : \hat{X} \longrightarrow X$$

an arbitrary fixed map fulfilling

$$\overline{\epsilon}(\mathcal{N}(x)) = x \qquad \forall x \in X,$$

and

$$\lambda C(\overline{\epsilon}(C)) \leq \epsilon \quad \forall C \in \hat{X}.$$

Now, if $C \in \hat{X}$ and $\Gamma \in 2^{\hat{X}}$, then we have

$$\begin{aligned}\delta_Y(f(\overline{\epsilon}(C)), f(\overline{\epsilon}(\Gamma))) &\leq \delta(\overline{\epsilon}(C), \overline{\epsilon}(\Gamma)) \\ &= \hat{\delta}(\mathcal{N}(\overline{\epsilon}(C)), e_X(\overline{\epsilon}(\Gamma))) \\ &\leq \hat{\delta}(\mathcal{N}(\overline{\epsilon}(C)), \{C\}) + \hat{\delta}(C, \Gamma) + \sup_{C' \in \Gamma} \hat{\delta}(C', e_X(\overline{\epsilon}(\Gamma))).\end{aligned}$$

From the definition of $\overline{\epsilon}$ it follows that

$$\hat{\delta}(\mathcal{N}(\overline{\epsilon}(C)), \{C\}) \leq \epsilon$$

and

$$\sup_{C' \in \Gamma} \hat{\delta}(C', e_X(\overline{\epsilon}(\Gamma))) \leq \epsilon,$$

and consequently

$$\delta_Y(f(\overline{\epsilon}(C)), f(\overline{\epsilon}(\Gamma))) \leq \hat{\delta}(C, \Gamma) + 2\epsilon.$$

On the other hand we have that

$$\delta_Y\left(\hat{f}(\mathcal{C}), \hat{f}(\boldsymbol{\Gamma})\right) \leq \delta_Y\left(\hat{f}(\mathcal{C}), \{f(\overline{\epsilon}(\mathcal{C}))\}\right) + \delta_Y(f(\overline{\epsilon}(\mathcal{C})), f(\overline{\epsilon}(\boldsymbol{\Gamma}))) + \sup_{\mathcal{C}' \in \boldsymbol{\Gamma}} \delta_Y\left(f(\overline{\epsilon}(\mathcal{C}')), \hat{f}(\boldsymbol{\Gamma})\right)$$

It follows from 6.3.5 and 7.1.10 that, for any $\mathcal{C}' \in \hat{X}$,

$$\begin{aligned}\delta_Y\left(\hat{f}(\mathcal{C}'), \{f(\overline{\epsilon}(\mathcal{C}'))\}\right) &= \lambda_Y f(\mathcal{C}')(f(\overline{\epsilon}(\mathcal{C}'))) \\ &\leq \lambda \mathcal{C}'(\overline{\epsilon}(\mathcal{C}')) \\ &\leq \epsilon.\end{aligned}$$

Combining the foregoing inequalities we thus obtain that

$$\begin{aligned}\delta_Y\left(\hat{f}(\mathcal{C}), \hat{f}(\boldsymbol{\Gamma})\right) &\leq \delta_Y(f(\overline{\epsilon}(\mathcal{C})), f(\overline{\epsilon}(\boldsymbol{\Gamma}))) + 2\epsilon \\ &\leq \hat{\delta}(\mathcal{C}, \boldsymbol{\Gamma}) + 4\epsilon,\end{aligned}$$

which by the arbitrariness of ϵ shows that $\hat{f}$ is indeed a contraction.

□

7.1.16 Corollary *The subconstruct* c**UAP**$_2$ *of* **AP** *is closed under the formation of products and closed subspaces in* **AP**.

Proof This is a consequence of 7.1.14 and 7.1.15.

□

7.1.17 Corollary *If* $(Y, \delta_Y) \in |\mathbf{cUAP}_2|$, $(X, \delta_X) \in |\mathbf{UAP}_2|$, $Z \subset X$ *is dense with respect to the topology underlying the* p**MET**$^\infty$*-bicoreflection of* (X, δ_X), *and* $f : (Z, \delta_{X|Z}) \longrightarrow (Y, \delta_Y)$ *is a contraction, then there exists a unique extension*

$$\overline{f} : (X, \delta_X) \longrightarrow (Y, \delta_Y)$$

which is also a contraction.

Proof The proof goes along exactly the same lines as the proof of the epireflectivity in 7.1.15 and we leave this verification to the reader.

□

7.1.18 Proposition *If* $(X,\delta) \in |\mathbf{UAP}_2|$, $(X_i,\delta_i) \in |\mathbf{cUAP}_2|$ *for* $i \in \{1,2\}$, *and*

$$e_i : (X,\delta) \longrightarrow (X_i,\delta_i) \qquad i \in \{1,2\}$$

are embeddings which are dense with respect to the topology underlying the $\mathbf{pMET}^\infty$*-bicoreflection of* (X,δ), *then* (X_1,δ_1) *and* (X_2,δ_2) *are isomorphic.*

Proof It follows from 7.1.17 that there exist unique contractions

$$\varphi_1 : (X_2,\delta_2) \longrightarrow (X_1,\delta_1) \quad \text{and} \quad \varphi_2 : (X_1,\delta_1) \longrightarrow (X_2,\delta_2),$$

such that $e_1 = \varphi_1 \circ e_2$ and $e_2 = \varphi_2 \circ e_1$. Since e_1 and e_2 are dense embeddings also with respect to the $\mathbf{TOP}$-bicoreflection of (X,δ) they are epimorphisms in $\mathbf{UAP}_2$ and it thus follows that $\varphi_1 \circ \varphi_2 = id_{X_1}$ and $\varphi_2 \circ \varphi_1 = id_{X_2}$.

□

7.1.19 Definition Given $(X,\delta) \in |\mathbf{UAP}_2|$ the space $\left(\hat{X},\hat{\delta}\right)$ is called the *completion of* (X,δ).

7.2 EXAMPLE: FUNCTION SPACES

For what follows in this section we refer to the results concerning function spaces given in 6.3 on completeness in function spaces.

7.2.1 Theorem *If* X *is a set,* (Y,δ_d) *is an* ∞p*-metric approach space, and* $\mathcal{Z} \subset Y^X$ *is endowed with the restriction* δ *of the product approach structure on* Y^X, *then the* $\mathbf{pMET}^\infty$*-bicoreflection of the completion of* $(\mathcal{Z},\delta)$ *is isomorphic to the usual metric completion of* $(\mathcal{Z},d_\delta)$.

Proof Let $\mathcal{D}$ be a symmetric gauge basis which generates δ. Further let $\left(\tilde{\mathcal{Z}},\tilde{d_\delta}\right)$ stand for the usual metric completion of $(\mathcal{Z},d_\delta)$, where $\tilde{\mathcal{Z}}$ stands for the set of minimal Cauchy filters in $\mathcal{Z}$. Clearly, if $\mathbf{H} \in \tilde{\mathcal{Z}}$, then $\mathbf{H}^{\mathcal{D}} \in \hat{\mathcal{Z}}$. Thus we obtain a map

$$\psi : \tilde{\mathcal{Z}} \longrightarrow \hat{\mathcal{Z}} : \mathbf{H} \longrightarrow \mathbf{H}^{\mathcal{D}}.$$

This map is onto. Indeed, if $\mathbf{F} \in \hat{\mathcal{Z}}$, then $\mathbf{F}$ is a $\mathcal{U}(\mathcal{D})$-Cauchy filter, i.e. for all $x \in X$, we have that

$$\mathbf{F}(x) \doteqdot \{\mathcal{A}(x) \mid \mathcal{A} \in \mathbf{F}\},$$

where $\mathcal{A}(x) \doteqdot \{f(x) \mid f \in \mathcal{A}\}$ is Cauchy in (Y, d). Since Y is complete, there exists $f \in Y^X$ such that, for all $x \in X$, $\mathbf{F}(x) \longrightarrow f(x)$. This implies that

$$\forall n \in \mathbb{N}_0, \forall x \in X\ \exists \mathcal{A}_0 \in \mathbf{F}: \ \mathcal{A}_0(x) \subset B_d\left(f(x), \frac{1}{2n}\right).$$

Also, since $\mathbf{F}$ is δ-Cauchy, we have that

$$\forall n \in \mathbb{N}_0\ \exists g_n \in \mathcal{Z}\ \forall x \in X: \left\{h \in \mathcal{Z} \mid d(g_n(x), h(x)) < \frac{1}{2n}\right\} \in \mathbf{F}.$$

Consequently, for all $n \in \mathbb{N}_0$ and for all $x \in X$, we have

$$\left\{h \in \mathcal{Z} \mid d(g_n(x), h(x)) < \frac{1}{2n}\right\} \cap \mathcal{A}_0 \neq \emptyset,$$

and thus, for all $x \in X$, $d(g_n(x), f(x)) \leq \frac{1}{n}$, i.e. $(g_n)_n \to f$ uniformly. Let $\mathbf{H}'$ stand for the filter generated by the sequence $(g_n)_n$ and let $\mathbf{H}$ be the minimal d_δ-Cauchy filter contained in $\mathbf{H}'$. Since both

$$\lambda_\delta \mathbf{F}(f) = 0 \quad \text{and} \quad \lambda_\delta \mathbf{H}(f) = 0,$$

$\mathbf{F}$ and $\mathbf{H}$ are $\mathcal{U}(\mathcal{D})$-equivalent. Since $\mathbf{F}$ is $\mathcal{U}(\mathcal{D})$-minimal this implies that $\psi(\mathbf{H}) = \mathbf{H}^{\mathcal{D}} = \mathbf{F}$.

Now let $\mathbf{F}, \mathbf{H} \in \hat{\mathcal{Z}}$; then we have

$$\begin{aligned}
\tilde{d}_\delta(\mathbf{F}, \mathbf{H}) &= \sup_{\mathcal{F} \in \mathbf{F}} \sup_{\mathcal{H} \in \mathbf{H}} \inf_{f \in \mathcal{F}} \inf_{h \in \mathcal{H}} \sup_{x \in X} d(f(x), h(x)) \\
&\geq \sup_{x \in X} \hat{d}_{\{x\}}\left(\mathbf{F}^{\mathcal{D}}, \mathbf{H}^{\mathcal{D}}\right) \\
&= d_{\hat{\delta}}\left(\mathbf{F}^{\mathcal{D}}, \mathbf{H}^{\mathcal{D}}\right).
\end{aligned}$$

Conversely, if

$$d_{\hat{\delta}}\left(\mathbf{F}^{\mathcal{D}}, \mathbf{H}^{\mathcal{D}}\right) < \epsilon,$$

then, given $\theta > 0$, we can find $f, h \in \mathcal{Z}$ such that

$$\lambda_{d_\delta} \mathbf{F}(f) < \frac{\theta}{2} \quad \text{and} \quad \lambda_{d_\delta} \mathbf{H}(f) < \frac{\theta}{2},$$

which implies that, for any $x \in X$,

$$B_{d_{\{x\}}}(f,\theta) \in \mathbf{F}^{\mathcal{D}} \quad \text{and} \quad B_{d_{\{x\}}}(h,\theta) \in \mathbf{H}^{\mathcal{D}}$$

It then follows that, for any $x \in X$, $d_{\{x\}}(f,h) \leq \epsilon+2\theta$ and thus $d_{\delta}(f,h) \leq \epsilon+2\theta$. Consequently,

$$\begin{aligned} \lambda_{d_\delta}\mathbf{H}(f) &\leq \lambda_{d_\delta}\mathbf{H}(h) + d_\delta(h,f) \\ &\leq \epsilon + 3\theta, \end{aligned}$$

which by the arbitrariness of θ finally implies that

$$\begin{aligned} \tilde{d}_\delta(\mathbf{F},\mathbf{H}) &= \inf_{f \in \mathcal{Z}} (\lambda_{d_\delta}\mathbf{F}(f) + \lambda_{d_\delta}\mathbf{H}(f)) \\ &\leq \epsilon. \end{aligned}$$

This proves at the same time that ψ is injective and an isomorphism and we are done.

□

8
COMPACTIFICATION

It is well known that compactifications do not often preserve metrizability. Thus for instance a nontrivial Čech–Stone compactification is never metrizable. And even if a compactification is metrizable, then it need not be metrizable by an extension of the given metric. This is the case for instance with the Alexandroff compactification of $\mathbb{R}$. In **UAP** the situation is better. We will construct a compactification which is the **AP**-counterpart for the Čech–Stone compactification in **TOP**. We will prove, moreover, that, in **TOP**, our compactification coincides with the Čech–Stone compactification, and that in general the **TOP**-coreflection of our compactification is the Smirnov compactification of an associated proximity. In the final section we then apply the general theory to prove that $\beta\mathrm{N}$ can be endowed with a distance which extends the usual metric on N, and which has the Čech–Stone compactification as topological coreflection; in other words, which "distancizes" the Čech–Stone topology of $\beta\mathrm{N}$.

8.1 CONSTRUCTION

We recall from the fourth chapter that, given an approach space (X, δ), we denote by $\mathcal{K}^*(X)$ the set of all bounded contractions from (X, δ) to $(\mathbb{R}, \delta_{d_E})$. Further, given any set of contractions $\mathcal{F} \subset \mathcal{K}^*(X)$, $\mathcal{D}_{\mathcal{F}}$ stands for the set of all p-metrics

$$d_{\mathcal{F}_0} : X \times X \longrightarrow [0, \infty] : (x, y) \longrightarrow \sup_{f \in \mathcal{F}_0} |f(x) - f(y)|,$$

where $\mathcal{F}_0$ ranges over the finite subsets of $\mathcal{F}$.

A set $\mathcal{D} \subset pq\mathrm{M}^\infty(X)$ of ∞pq-metrics (respectively a set of contractions $\mathcal{F} \subset \mathcal{K}^*(X)$) is said to be *point-separating* or to *separate points* if, for any $x, y \in X$, $x \neq y$, there exists $d \in \mathcal{D}$ (respectively $f \in \mathcal{F}$) such that $d(x, y) \neq 0$ (respectively $f(x) \neq f(y)$).

Similarly to 4.1.6 and 4.1.11, we then obtain the following characterizations of the objects in $\mathbf{UAP}_2$.

8.1.1 Proposition *For a uniform approach space (X, δ) the following are equivalent:*

1. *(X, δ) is a Hausdorff uniform approach space, i.e. it is an object in $\mathbf{UAP}_2$,*
2. *$\mathcal{G}$ is point-separating,*
3. *every set $\mathcal{F} \subset \mathcal{K}^*(X)$ which generates δ is point-separating,*
4. *$\mathcal{K}^*(X)$ is point-separating,*
5. *(X, δ) is isomorphic to a subspace of a power of $\mathbb{R}$ (equipped with δ_{d_E}).*

Proof $1 \Rightarrow 2$. Since $(X, \mathcal{T}_\delta)$ is Hausdorff, for any $x, y \in X$, $x \neq y$, we can find $d, d' \in \mathcal{G}$, $\epsilon > 0$, and $\epsilon' > 0$ such that

$$\{d(x, \cdot) < \epsilon\} \cap \{d'(y, \cdot) < \epsilon'\} = \emptyset.$$

Hence, if we put $d'' \doteqdot d \vee d'$, then $d''(x, y) > 0$.

$2 \Rightarrow 3$. If $\mathcal{F} \subset \mathcal{K}^*(X)$ generates δ, then it follows from 1.8.8 that $\mathcal{D}_\mathcal{F}$ is a basis for $\mathcal{G}$. Hence, given $x, y \in X$, $x \neq y$, there exists $\mathcal{F}_0 \in 2^{(\mathcal{F})}$ such that $d_{\mathcal{F}_0}(x, y) > 0$ from which it follows that there exists $f \in \mathcal{F}_0$ such that $f(x) \neq f(y)$.

$3 \Rightarrow 4$. This is trivial.

$4 \Rightarrow 5$. Consider the product space $\mathbb{R}^{\mathcal{K}^*(X)}$ where each copy of $\mathbb{R}$ is endowed with δ_{d_E}. Then it follows from the fact that $\mathcal{K}^*(X)$ is point-separating that the map

$$e : X \longrightarrow \mathbb{R}^{\mathcal{K}^*(X)} : x \longrightarrow (f(x))_{f \in \mathcal{K}^*(X)},$$

is injective and hence an embedding.

$5 \Rightarrow 1$. This follows at once from the fact that $\mathbb{R}$, endowed with δ_{d_E}, is a Hausdorff uniform approach space and from 4.3.2.

□

8.1.2 Proposition *For a Hausdorff uniform approach space (X, δ) the following are equivalent:*

1. (X, δ) is compact,

2. (X, δ) is isomorphic to a closed subspace of a product of compact subsets of $\mathbb{R}$ (each endowed with the restriction of δ_{d_E}).

Proof That the second property implies the first one is a consequence of the fact that an isomorphism between approach spaces induces a homeomorphism between their topological bicoreflections. To see that the first property implies the second one, let (X, δ) be a Hausdorff compact uniform approach space. Then it follows from 8.1.1 and the fact that, for each $f \in \mathcal{K}^*(X)$, $f(X) \subset \mathbb{R}$ is compact, that

$$e : X \longrightarrow \prod_{f \in \mathcal{K}^*(X)} f(X) \subset \mathbb{R}^{\mathcal{K}^*(X)} : x \longrightarrow (f(x))_{f \in \mathcal{K}^*(X)}$$

is an embedding into a product of compact subsets of $\mathbb{R}$. Since X itself is compact, it follows that the embedding is also closed.

□

In what follows we denote by $k\mathbf{UAP}_2$ the full and isomorphism-closed subconstruct of $\mathbf{UAP}_2$ consisting of all Hausdorff compact uniform approach spaces.

8.1.3 Corollary *$k\mathbf{UAP}_2$ is an epireflective subconstruct of $\mathbf{UAP}_2$.*

Proof This follows at once from 7.1.14, 8.1.2, and Adámek *et al.* [1].

□

We recall (see 4.1.5) that, for a given approach space (X, δ),

$$\mathcal{G}^{\mathrm{s}} = \{d \in p\mathrm{M}^{\infty}(X) \mid \delta_d \leq \delta\}.$$

The subset of all bounded ∞p-metrics in $\mathcal{G}^{\mathrm{s}}$ will be denoted by $\mathcal{G}^{\mathrm{s}}_b$. Clearly $\mathcal{G}^{\mathrm{s}}_b$ is closed under the formation of finite suprema and it is the largest collection of bounded p-metrics which generates δ. Obviously too, $\mathcal{U}(\delta) = \mathcal{U}(\mathcal{G}^{\mathrm{s}}_b)$.

8.1.4 Definition A ($\mathbf{UAP}_2$-)*compactification* of a Hausdorff uniform approach space (X, δ) is a pair $((X', \delta'), e)$, where (X', δ') is an object in $k\mathbf{UAP}_2$, i.e. a Hausdorff compact uniform approach space, and where

$$e : (X, \delta) \longrightarrow (X', \delta')$$

is a dense embedding. Usually we will refer to (X', δ') as being a compactification of (X, δ).

Each Hausdorff uniform approach space has a compactification. Consider a subset $\mathcal{F}$ of $\mathcal{K}^*(X)$ which generates δ. For each $f \in \mathcal{F}$, let I_f be a compact subset of $\mathbb{R}$ containing the image of f. Let

$$(X', \delta_{\mathcal{F}}) \doteqdot \left(\prod_{f \in \mathcal{F}} I_f, \prod_{f \in \mathcal{F}} \delta_{d_E} \right)$$

Then

$$e_{\mathcal{F}} : (X, \delta) \longrightarrow (X', \delta_{\mathcal{F}}) : x \longrightarrow (f(x))_{f \in \mathcal{F}}$$

is an embedding and, obviously, if we let $\beta_{\mathcal{F}}$ be the same map as $e_{\mathcal{F}}$ but with range $\mathrm{cl}_{\mathcal{T}_{\delta_{\mathcal{F}}}}(e_{\mathcal{F}}(X))$, then $\left(\left(\mathrm{cl}_{\mathcal{T}_{\delta_{\mathcal{F}}}}(e_{\mathcal{F}}(X)), \delta_{\mathcal{F}}\right), \beta_{\mathcal{F}}\right)$ is a compactification of (X, δ). We will denote this compactification by $(\beta_{\mathcal{F}} X, \delta_{\mathcal{F}})$ or by $\beta_{\mathcal{F}} X$ for short. The appropriateness of this notation will become clear in the sequel.

Notice that $\delta_{\mathcal{F}}$ is generated by the symmetric gauge basis

$$\mathcal{D}^*_{\mathcal{F}} \doteqdot \left\{ \sup_{f \in \mathcal{H}} d_f \mid \mathcal{H} \in 2^{(\mathcal{F})} \right\},$$

where for each $f \in \mathcal{F}$

$$d^*_f : \beta_{\mathcal{F}} X \times \beta_{\mathcal{F}} X \longrightarrow [0, \infty] : (p, q) \longrightarrow \left| pr_f(p) - pr_f(q) \right|.$$

Notice also that if (X, δ) is compact, then $\beta_{\mathcal{F}}$ is an isomorphism.

In order not to overload the notation, when we construct a compactification as above, starting from the entire set of contractions $\mathcal{K}^*(X)$, we will drop all subscripts and simply put $\beta^* X \doteqdot \beta_{\mathcal{K}^*(X)} X$, $\beta^* \delta \doteqdot \delta_{\mathcal{K}^*(X)}$, and $\beta^* \doteqdot \beta_{\mathcal{K}^*(X)}$.

8.1.5 Theorem *For any Hausdorff uniform approach space (X, δ) the compactification $\beta^* X$ has the following equivalent properties:*

1. every $f \in \mathcal{K}^*(X)$ has a unique extension $f^* \in \mathcal{K}^*(\beta^* X)$, i.e. there exists a unique bounded contraction f^* such that the diagram

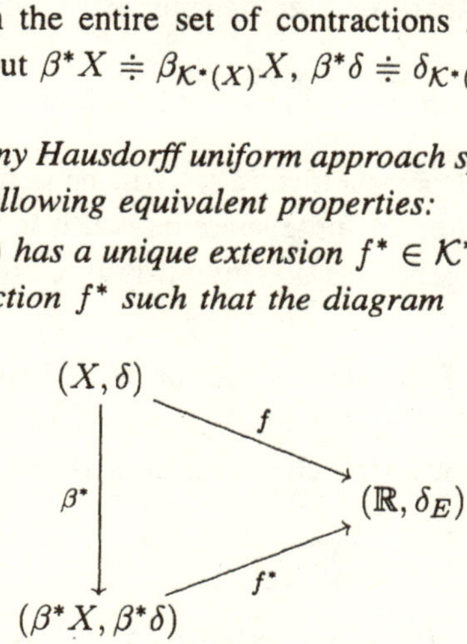

commutes,

2. *for any Hausdorff compact uniform approach space* (X', δ') *and any contraction* $f : (X, \delta) \longrightarrow (X', \delta')$, *there exists a unique extension to* $\beta^* X$, *i.e. there exists a unique contraction* $f^* : (\beta^* X, \beta^* \delta) \longrightarrow (X', \delta')$ *such that the diagram*

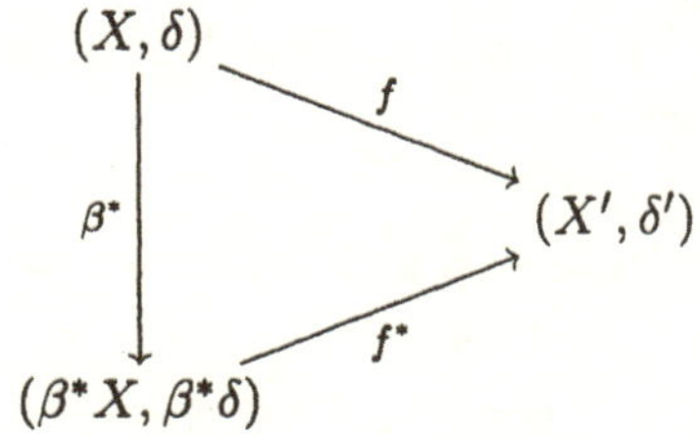

commutes.

This compactification is essentially unique in the sense that any other compactification with the same properties is isomorphic to $\beta^* X$ *under an isomorphism which leaves* X *pointwise fixed.*

Proof The proof goes in the same way as the analogous proof for the Čech–Stone compactification in **TOP**. For example, if $f \in \mathcal{K}^*(X)$, then $f^* \in \mathcal{K}^*(\beta^*(X))$ is defined as the restriction of the projection pr_f to $\beta^* X$. If h is another such extension, then it suffices to notice that $\{x \in \beta^* X \mid h(x) = f^*(x)\}$ is a closed subset of $\beta^* X$ containing $e_{\mathcal{K}^*(X)}(X)$.

To show that $\beta^* X$ is essentially unique, consider for instance two compactifications $((X_1, \delta_1), e_1)$ and $((X_2, \delta_2), e_2)$ fulfilling the second property. Then we can find contractions $\psi_1 : (X_2, \delta_2) \longrightarrow (X_1, \delta_1)$ and $\psi_2 : (X_1, \delta_1) \longrightarrow (X_2, \delta_2)$ such that $e_1 = \psi_1 \circ e_2$ and $e_2 = \psi_2 \circ e_1$. Consequently, ψ_1 and ψ_2 are isomorphisms. □

The following result is an immediate consequence of 8.1.5.

8.1.6 Corollary *For any* $(X, \delta) \in |\mathbf{UAP_2}|$, *the map*

$$\beta^* : (X, \delta) \longrightarrow (\beta^* X, \beta^* \delta)$$

is an epireflection of (X, δ) *in* $k\mathbf{UAP_2}$.

□

Notice that, in the same way as the Čech–Stone compactification, $\beta^* X$ is the largest compactification of (X, δ). For any other compactification $((X', \delta'), e)$

there exists a contraction

$$c : (\beta^* X, \beta^* \delta) \longrightarrow (X', \delta')$$

such that $e = c \circ \beta^*$.

8.1.7 Theorem *If (X, δ) is a Tychonoff topological approach space, then $\beta^* X$ is the Čech–Stone compactification.*

Proof By our previous results, it is sufficient to show that $\beta^* X$ is topological. Let $x \in \beta^* X$ and $A \subset \beta^* X$; then

$$\beta^* \delta(x, A) = \sup_{\mathcal{H} \in 2^{(\mathcal{K}^*(X))}} \inf_{a \in A} \sup_{f \in \mathcal{H}} \left| \mathrm{pr}_f(x) - \mathrm{pr}_f(a) \right|.$$

Since X is topological we know that $\mathcal{K}^*(X) = C^*(X)$, where $C^*(X)$ is the set of all bounded continuous real-valued functions on X. Consequently, if $f \in \mathcal{K}^*(X)$ and $\alpha \in \mathbb{R}$, then $\alpha f \in \mathcal{K}^*(X)$. Moreover, since both $\mathrm{pr}_{\alpha f}$ and $\alpha \mathrm{pr}_f$ are contractions which extend αf to $\beta^* X$, we also have that $\alpha \mathrm{pr}_f = \mathrm{pr}_{\alpha f}$. Now, if $\beta^* \delta(x, A) > 0$, then there exists $\mathcal{H} \in 2^{(\mathcal{K}^*(X))}$ such that

$$\inf_{a \in A} \sup_{f \in \mathcal{H}} \left| \mathrm{pr}_f(x) - \mathrm{pr}_f(a) \right| > 0,$$

and then it follows that

$$\begin{aligned} \beta^* \delta(x, A) &= \sup_{\mathcal{H} \in 2^{(\mathcal{K}^*(X))}} \inf_{a \in A} \sup_{f \in \mathcal{H}} \left| \mathrm{pr}_f(x) - \mathrm{pr}_f(a) \right| \\ &\geq \sup_{\alpha \geq 1} \inf_{a \in A} \sup_{f \in \mathcal{H}} \left| \mathrm{pr}_{\alpha f}(x) - \mathrm{pr}_{\alpha f}(a) \right| \\ &= \sup_{\alpha \geq 1} \inf_{a \in A} \sup_{f \in \mathcal{H}} \alpha \left| \mathrm{pr}_f(x) - \mathrm{pr}_f(a) \right| \\ &= \infty. \end{aligned}$$

Thus $\beta^* \delta$ attains only the values 0 and ∞ which, by 2.1.9, proves that $\beta^* X$ is topological.

□

It is clear that the topological bicoreflection of $\beta^* X$ is a compactification (in the topological sense) of the topological bicoreflection of X. In general it is smaller than the Čech–Stone compactification. We will now proceed to give a more concrete characterization of $\beta^* X$ which will clarify the situation. In this description we will make use of proximity spaces and related concepts. We refer the reader to Naimpally and Warrack [94] for an account of the basic concepts of proximity spaces which we require (see also Appendix B).

The following result gives a basic relationship which we will need.

8.1.8 Proposition *If (X,δ) is a uniform approach space, generated by the symmetric gauge basis $\mathcal{D}$, then the relation $\Delta_{\mathcal{D}}$ on $2^X \times 2^X$, defined by*

$$A\Delta_{\mathcal{D}}B \quad \Leftrightarrow \quad \sup_{d\in\mathcal{D}} \inf_{a\in A} \inf_{b\in B} d(a,b) = 0,$$

is a proximity relation which is compatible with the topological bicoreflection of (X,δ), i.e. for any $x \in X$ and $A \subset X$, we have

$$\{x\}\Delta_{\mathcal{D}}A \quad \Leftrightarrow \quad \delta(x,A) = 0.$$

Proof This is straightforward and is left to the reader.

□

8.1.9 Example The same example as the third one in 4.4.4 proves that different symmetric gauges can generate the same distance but different proximities. Again consider the real line $\mathbb{R}$ and the two collections of ∞p-metrics $\mathcal{D} \doteqdot \{d_a \mid a > 0\}$ and $\mathcal{D}' \doteqdot \{d_E\}$, where for each real number $a > 0$, $f_a \doteqdot (id_{\mathbb{R}} \wedge a) \vee (-a)$, where

$$d_a(x,y) \doteqdot |f_a(x) - f_a(y)| \qquad \forall x,y \in \mathbb{R},\ \forall a > 0$$

and where d_E stands for the Euclidean metric. Then $\mathcal{D}$ and $\mathcal{D}'$ are bases for the same symmetric gauge, but $\Delta_{\mathcal{D}} \neq \Delta_{\mathcal{D}'}$.

We recall that, given a proximity space (X,Δ), a set p of subsets of X is called a *cluster* if the following conditions are fulfilled:

1. $\forall A,B \in p : \ A\Delta B$,
2. if $A\Delta B$ for every $B \in p$ then $A \in p$,
3. if $A \cup B \in p$ then $A \in p$ or $B \in p$.

Given $(X,\delta) \in |\mathbf{UAP_2}|$ and a subset $\mathcal{F} \subset \mathcal{K}^*(X)$ which generates (X,δ), we will denote the set of all clusters in the proximity space $(X,\Delta_{\mathcal{D}_\mathcal{F}})$ by $\kappa_\mathcal{F}X$. Further we let

$$c_\mathcal{F} : X \longrightarrow \kappa_\mathcal{F}X : x \longrightarrow \{A \subset X \mid \delta(x,A) = 0\}$$

be the canonical injection (see e.g. Naimpally and Warrack [94]). We identify the points of X with their images under the map $c_\mathcal{F}$. The *Smirnov compactification* of $(X,\Delta_{\mathcal{D}_\mathcal{F}})$ is the set $\kappa_\mathcal{F}X$ equipped with the proximity relation for which two subsets $\mathcal{A}$ and $\mathcal{B}$ of $\kappa_\mathcal{F}X$ are proximal if any subsets A and B of X which absorb

$\mathcal{A}$ and $\mathcal{B}$ respectively are proximal in X. We recall that A is said to *absorb* $\mathcal{A}$ if, for all $q \in \mathcal{A}$, we have that $A \in q$, i.e. if $A \in \bigcap_{q \in \mathcal{A}} q$. Again we refer to Naimpally and Warrack [94] for more details.

For any $A \in 2^X \setminus \{\emptyset\}$, we now define

$$\hat{\delta}_A : \kappa_{\mathcal{F}} X \longrightarrow [0, \infty] : p \longrightarrow \inf \left\{ \epsilon \geq 0 \mid \forall d \in \mathcal{D}_{\mathcal{F}} : A_d^{(\epsilon)} \in p \right\},$$

where, for any $d \in \mathcal{D}_{\mathcal{F}}$ and $\epsilon \geq 0$,

$$A_d^{(\epsilon)} \doteq \{x \in X \mid \delta_d(x, A) \leq \epsilon\}.$$

Then we define

$$\delta^*_{\mathcal{F}} : \kappa_{\mathcal{F}} X \times \kappa_{\mathcal{F}} X \longrightarrow [0, \infty]$$

as follows. For any $p \in \kappa_{\mathcal{F}} X$ and $\mathcal{A} \subset \kappa_{\mathcal{F}} X$,

$$\delta^*_{\mathcal{F}}(p, \mathcal{A}) \doteq \begin{cases} \infty & \mathcal{A} = \emptyset, \\ \sup \left\{ \hat{\delta}_A(p) \mid A \text{ absorbs } \mathcal{A} \right\} & \mathcal{A} \neq \emptyset. \end{cases}$$

Further, if $A \subset X$ and $\epsilon \geq 0$, then, as usual, we will put $A^{(\epsilon)} = \{x \in X \mid \delta(x, A) \leq \epsilon\}$ and if we consider A as a subset of $\kappa_{\mathcal{F}} X$ then we will put

$$A^{(\epsilon)^*} \doteq \{p \in \kappa_{\mathcal{F}} X \mid \delta^*_{\mathcal{F}}(p, A) \leq \epsilon\}.$$

8.1.10 Lemma *Let (X, δ) be a uniform approach space, generated by the symmetric gauge basis $\mathcal{D}$. Then, for any $A \subset X$ and any cluster $p \in \kappa_{\mathcal{F}} X$, the following are equivalent:*

1. *$A \in p$,*
2. *$\forall d \in \mathcal{D}, \forall \epsilon > 0 : \; A_d^{(\epsilon)} \in p$.*

Proof We only need to prove that the second condition implies the first one. Suppose that $A \notin p$; then there exists $C \in p$ such that $A \neg \Delta_{\mathcal{D}} C$. Hence, there also exists $d \in \mathcal{D}$ and $\epsilon > 0$ such that $A_d^{(\epsilon)} \cap C_d^{(\epsilon)} = \emptyset$, which implies that $A_d^{(\epsilon)} \notin p$. □

8.1.11 Theorem *If (X, δ) is a Hausdorff uniform approach space, then the following hold:*

1. *$(\kappa_{\mathcal{F}} X, \delta^*_{\mathcal{F}}) \in |\mathbf{AP}|$,*

2. $c_{\mathcal{F}} : (X, \delta) \longrightarrow \left(\kappa_{\mathcal{F}} X, \delta^*_{\mathcal{F}}\right)$ is a dense embedding,

3. the topological bicoreflection of $\left(\kappa_{\mathcal{F}} X, \delta^*_{\mathcal{F}}\right)$ is (homeomorphic to) the topological space underlying the Smirnov compactification of $(X, \Delta_{\mathcal{D}_{\mathcal{F}}})$.

Proof 1. First of all note that, for all $A, B \subset X$, we have that $A \in \bigcap \{q \mid q \in B\}$ if and only if $B \subset \overline{A}$. From this we conclude that

$$\begin{aligned}\delta^*_{\mathcal{F}}(p, B) &= \sup \left\{ \hat{\delta}_A(p) \mid A \in \bigcap_{q \in B} q \right\} \\ &= \hat{\delta}_B(p).\end{aligned}$$

It is clear that $\delta^*_{\mathcal{F}}$ satisfies (D1) and (D2). Moreover, if $p \in \kappa_{\mathcal{F}} X$ and $\mathcal{A} \subset \mathcal{A}' \subset \kappa_{\mathcal{F}} X$, then it is clear from the definition that $\delta^*_{\mathcal{F}}(p, \mathcal{A}') \leq \delta^*_{\mathcal{F}}(p, \mathcal{A})$. Consequently, in order to prove (D3), only one inequality has to be shown. Now, if $A, B \subset X$, then we have

$$\delta^*_{\mathcal{F}}(p, A \cup B) = \delta^*_{\mathcal{F}}(p, A) \wedge \delta^*_{\mathcal{F}}(p, B),$$

and the general case follows from this since, if A absorbs $\mathcal{A}$ and B absorbs $\mathcal{B}$, then $A \cup B$ absorbs $\mathcal{A} \cup \mathcal{B}$. To prove also that (D4) is satisfied, we first make some preliminary observations.

Claim 1: For any $\alpha, \beta \in \mathbb{R}^+$, $d \in \mathcal{D}_{\mathcal{F}}$, and $A \subset X$, we have $\left(A_d^{(\alpha)}\right)_d^{(\beta)} \subset A_d^{(\alpha+\beta)}$

Indeed, this is nothing more than (T4) for δ_d.

Claim 2: For any $A \subset X$ and $\epsilon > 0$, we have

$$A^{(\epsilon)^*} = \bigcap_{\epsilon' > \epsilon} \left\{ q \mid \forall d \in \mathcal{D}_{\mathcal{F}} : A_d^{(\epsilon')} \in q \right\}.$$

Indeed, this follows from the definitions of $\delta^*_{\mathcal{F}}$ and $\hat{\delta}_A$.

Claim 3: If A absorbs $\mathcal{A}$, then for any $\epsilon > 0$ we have $\mathcal{A}^{(\epsilon)} \subset A^{(\epsilon)^*}$.

Indeed, if $p \in \mathcal{A}^{(\epsilon)}$ then it follows from the second claim that, for all $\epsilon' > \epsilon$, for all $d \in \mathcal{D}_{\mathcal{F}}$, and for any B which absorbs $\mathcal{A}$, we have $B_d^{(\epsilon')} \in p$. Since A absorbs $\mathcal{A}$ it follows again from the second claim that $p \in A^{(\epsilon)^*}$.

Now, consider first the case where $p \in \kappa_{\mathcal{F}} X$ and where $A \subset X$. From the first and second claims we can then deduce that, for any $\epsilon \geq 0$,

$$\hat{\delta}_A(p) \leq \delta^*_{\mathcal{F}}\left(p, A^{(\epsilon)^*}\right) + \epsilon.$$

For the general case, let $\mathcal{A} \subset \kappa_{\mathcal{F}}X$; then it follows from the foregoing inequality that

$$\delta^*_{\mathcal{F}}(p, \mathcal{A}) \leq \sup\left\{\delta^*_{\mathcal{F}}\left(p, A^{(\epsilon)^*}\right) \mid A \text{ absorbs } \mathcal{A}\right\} + \epsilon$$

and the result follows from the third claim.

This proves that $(\kappa_{\mathcal{F}}X, \delta^*_{\mathcal{F}})$ is indeed an approach space.

2. That $c_{\mathcal{F}}$ is an embedding follows from the fact that, if $x \in X$ and $A \subset X$, then

$$\begin{aligned}\delta^*_{\mathcal{F}}(x, A) &= \inf\left\{\epsilon \geq 0 \mid \forall d \in \mathcal{D}_{\mathcal{F}} : x \in A_d^{(\epsilon)}\right\} \\ &= \delta(x, A).\end{aligned}$$

That the embedding is dense follows from the fact that X belongs to any cluster $p \in \kappa_{\mathcal{F}}X$.

3. Given $p \in \kappa_{\mathcal{F}}X$ and $\mathcal{A} \subset \kappa_{\mathcal{F}}X$, p is in the closure of $\mathcal{A}$ for the topology of the Smirnov compactification if $A \in p$, for any A which absorbs $\mathcal{A}$. By the definition of $\delta^*_{\mathcal{F}}$, p is in the closure of $\mathcal{A}$ for the topological bicoreflection of $(\kappa_{\mathcal{F}}X, \delta^*_{\mathcal{F}})$ if $A_d^{(\epsilon)} \in p$, for any $\epsilon > 0$, any $d \in \mathcal{D}_{\mathcal{F}}$, and any A which absorbs $\mathcal{A}$. It follows from 8.1.10 that these two conditions are equivalent.

□

Before proceeding with the alternative description of our compactification we need another lemma.

8.1.12 Lemma *If $(X, \delta) \in |\mathbf{UAP}|$ and Y is a dense subspace of X, then, for any $x \in X$ and $A \subset X$, we have*

$$\delta(x, A) = \sup\left\{\delta(x, B) \mid B \subset Y,\ A \subset \overline{B}\right\}.$$

Proof Let $\mathcal{D}$ be a symmetric gauge for (X, δ). If $\delta(x, A) > \epsilon > 0$, then there exists $d_0 \in \mathcal{D}$ and $\theta > 0$ such that $\delta_{d_0}(x, A) > \epsilon + \theta$. If we let

$$B' \doteqdot \{y \in X \mid \delta_{d_0}(y, A) < \theta\},$$

and $B \doteqdot B' \cap Y$, then it follows from the fact that B' is open and that Y is dense that $A \subset \overline{B}$. Since further

$$\begin{aligned}\delta(x, B) + \theta &\geq \delta_{d_0}(x, B) + \theta \\ &\geq \delta_{d_0}(x, B) + \sup_{b \in B} \delta_{d_0}(b, A) \\ &\geq \delta_{d_0}(x, A) \\ &\geq \epsilon + \theta,\end{aligned}$$

this proves one inequality. As the other inequality is trivial, we are done.

□

We now define the map

$$\Psi : \kappa_{\mathcal{F}}X \longrightarrow \beta_{\mathcal{F}}X,$$

which assigns to each cluster $p \in \kappa_{\mathcal{F}}X$ the unique adherence point of the unique cluster in $\beta_{\mathcal{F}}X$ which contains p.

8.1.13 Theorem *If (X, δ) is a Hausdorff uniform approach space which is generated by the set of contractions $\mathcal{F} \subset \mathcal{K}^*(X)$, then*

$$\Psi : (\kappa_{\mathcal{F}}X, \delta^*_{\mathcal{F}}) \longrightarrow (\beta_{\mathcal{F}}X, \delta_{\mathcal{F}})$$

is an isomorphism.

Proof Clearly the map Ψ is well defined and bijective and the diagram

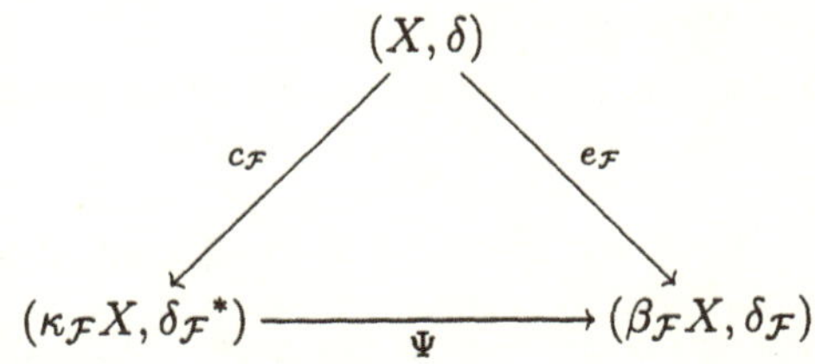

is commutative. Now first suppose that $p \in \kappa_{\mathcal{F}}X$ and that $A \subset X$. Then

$$\begin{aligned}
\delta_{\mathcal{F}}(\Psi(p), A) &= \sup_{\mathcal{H} \in 2^{(\mathcal{F})}} \inf_{a \in A} \sup_{f \in \mathcal{H}} \left|\mathrm{pr}_f(\Psi(p)) - \mathrm{pr}_f(a)\right| \\
&= \sup_{d^* \in \mathcal{D}^*_{\mathcal{F}}} \inf_{a \in A} d^*(\Psi(p), a) \\
&= \sup_{d^* \in \mathcal{D}^*_{\mathcal{F}}} \delta_{d^*}(\Psi(p), A) \\
&= \inf\left\{\epsilon \geq 0 \mid \forall d^* \in \mathcal{D}^*_{\mathcal{F}} : \Psi(p) \in A^{(\epsilon)}_{d^*}\right\} \\
&= \inf\left\{\epsilon \geq 0 \mid \forall d \in \mathcal{D}_{\mathcal{F}} : A^{(\epsilon)}_{d} \in p\right\} \\
&= \delta^*_{\mathcal{F}}(p, A).
\end{aligned}$$

Notice that the second-last equality follows from the fact that $\Psi(p) \in A^{(\epsilon)}_{d^*}$ implies that, for all $\epsilon' > \epsilon$, we have $A^{(\epsilon')}_{d} \in p$ and similarly that $A^{(\epsilon)}_{d} \in p$ implies that, for all $\epsilon' > \epsilon$, we have $\Psi(p) \in A_{d^*}^{(\epsilon')}$. It follows, in particular, that $p \in \overline{A}$ if and only if $\Psi(p) \in \overline{A}$. By definition of the Smirnov compactification it further follows that if $A \subset X$ and $\mathcal{A} \subset \kappa_{\mathcal{F}}X$, then A absorbs $\mathcal{A}$ if and only if $\Psi(\mathcal{A}) \subset \overline{A}$. So if $p \in \kappa_{\mathcal{F}}X$ and $\mathcal{A} \subset \kappa_{\mathcal{F}}X$ then, by definition of $\delta^*_{\mathcal{F}}$ and 8.1.12, we have

$$\begin{aligned}\delta^*_{\mathcal{F}}(p, \mathcal{A}) &= \sup \left\{ \delta^*_{\mathcal{F}}(p, A) \mid A \in \bigcap_{q \in \mathcal{A}} q \right\} \\ &= \sup \left\{ \delta_{\mathcal{F}}(\Psi(p), A) \mid \Psi(\mathcal{A}) \subset \overline{A} \right\} \\ &= \delta_{\mathcal{F}}(\Psi(p), \Psi(\mathcal{A})).\end{aligned}$$

This proves that Ψ is an isomorphism, and we are done.

□

Our description via clusters allows us to answer the question as to when the topological bicoreflection of our compactification coincides with the Čech–Stone compactification. Obviously we need to look only at the case where $\mathcal{F} = \mathcal{K}^*(X)$, i.e. we can restrict our attention to $\beta^* X$.

If X is a Tychonoff space, then the finest compatible proximity relation Δ_{fine} is given by

$$A \Delta_{fine} B \Leftrightarrow A \text{ and } B \text{ cannot be completely separated.}$$

This proximity relation is called the *fine proximity*. We recall that two sets A and B are said to be completely separated if there exists a continuous function $f : X \longrightarrow [0, 1]$ such that $f(A) = 0$ and $f(B) = 1$.

If X is normal then $A \Delta_{fine} B$ is equivalent with $\overline{A} \cap \overline{B} \neq \emptyset$. Let $\mathbf{Z}(X)$ stand for the set of all zero sets in X.

8.1.14 Theorem *Let (X, δ) be a Hausdorff uniform approach space. Then the topological bicoreflection of $\beta^* X$ is isomorphic to the Čech–Stone compactification of the topological bicoreflection of (X, δ) if and only if $\Delta_{\mathcal{D}_{\mathcal{K}^*(X)}}$ is the fine proximity.*

Proof We have the following equivalences:

$$\forall A, B \in \mathbf{Z}(X) : A \cap B = \emptyset \Rightarrow A \text{ and } B \text{ have disjoint closures in } \beta^* X$$

$$\Leftrightarrow$$

$$\forall A, B \in \mathbf{Z}(X) : A \cap B = \emptyset \Rightarrow \text{no } \Delta_{\mathcal{D}_{\mathcal{K}^*(X)}} \text{cluster contains both } A \text{ and } B$$

$$\Leftrightarrow$$

$$\forall A, B \in \mathbf{Z}(X) : A \cap B = \emptyset \;\Rightarrow\; A \text{ is not } \Delta_{\mathcal{D}_{\mathcal{K}^*(X)}} \text{ proximal to } B$$

$$\Leftrightarrow$$

$$\Delta_{\mathcal{D}_{\mathcal{K}^*(X)}} = \Delta_{fine}.$$

Now it suffices to notice that the Čech–Stone compactification βX is characterized by the property that disjoint zero sets in X have disjoint closures in βX.

□

We can now immediately derive a result which determines which ∞p-metric approach spaces have a compactification such that the Čech–Stone compactification of the underlying topological space is "distancized" by a distance which extends the ∞p-metric.

We recall that a metric space X is called a *UC space* or an *Atsuji space* (see e.g. Atsuji [4] and Chavez [24]) if any continuous real-valued function on X is uniformly continuous. Many equivalent characterizations of Atsuji spaces have been given in the literature. One of those characterizations which is of particular interest to us is the following: a space X is an Atsuji space if and only if, for any pair of disjoint closed subsets $A, B \subset X$, we have that $\inf_{a \in A} \inf_{b \in B} d(a, b) > 0$.

8.1.15 Corollary *For* $(X, d) \in |\mathbf{MET}^{\infty}|$ *the following are equivalent:*

1. the topological bicoreflection of $\beta^* X$ *coincides with the Čech–Stone compactification of* $(X, \mathcal{T}_d)$,

2. (X, d) *is an Atsuji space.*

Proof This follows immediately from 8.1.14.

□

8.2 EXAMPLE: THE CASE OF $\beta\mathbb{N}$

The last result of the foregoing section is applicable to $\mathbb{N}$ since, equipped with the usual Euclidean metric, $\mathbb{N}$ is clearly an Atsuji space. This implies that, although $\beta\mathbb{N}$ is not metrizable, it is in a natural way "distancizable". However, it implies much more, since the distance with which we can equip $\beta\mathbb{N}$ is actually an extension of the usual Euclidean metric on $\mathbb{N}$. From 8.1.15 it follows that the underlying set of $\beta\mathbb{N}$ is identical to the underlying set of $\beta^*\mathbb{N}$. We will therefore maintain the notation $\beta\mathbb{N}$. The distance on $\mathbb{N}$, generated by the Euclidean metric, is also simply denoted by δ_{d_E}. The distance $\beta^*\delta_{d_E}$ will be denoted by $\delta_{\beta\mathbb{N}}$ for short.

We recapitulate the properties this distance has in the following result.

8.2.1 Theorem *1. The distance $\delta_{\beta\mathbb{N}}$ on $\beta\mathbb{N}$ is given by*

$$\delta_{\beta\mathbb{N}}(\mathcal{F}, \mathbf{A}) \doteqdot \sup_{A \in \cap\{\mathcal{U} | \mathcal{U} \in \mathbf{A}\}} \sup_{F \in \mathcal{F}} \inf_{a \in A} \inf_{y \in F} |a - y|$$

for all $\mathcal{F} \in \beta\mathbb{N}$ and $\mathbf{A} \subset \beta\mathbb{N}$.

2. The restriction of $\delta_{\beta\mathbb{N}}$ to $\mathbb{N}$ coincides with the Euclidean distance δ_{d_E}.

3. The topological bicoreflection of $(\beta\mathbb{N}, \delta_{\beta\mathbb{N}})$ is $\beta\mathbb{N}$ (equipped with the topology of the Čech–Stone compactification).

4. The metric bicoreflection of $(\beta\mathbb{N}, \delta_{\beta\mathbb{N}})$ is such that the associated topology is discrete.

Proof The first three properties are easy consequences of the foregoing results and the fourth property actually follows from 8.2.2 where an explicit formula is derived for the metric bicoreflection.

□

That $\delta_{\beta\mathbb{N}}$ is not trivial on the remainder of $\beta\mathbb{N}$, $\mathbb{N}^* \doteqdot \beta\mathbb{N} \setminus \mathbb{N}$, is shown by the following result. Given an ultrafilter $\mathcal{U}$ on $\mathbb{N}$, we define

$$\mathcal{U} \oplus n \doteqdot \{U \oplus n \mid U \in \mathcal{U}\}$$

and

$$\mathcal{U} \ominus n \doteqdot \{U \ominus n \mid U \in \mathcal{U}\},$$

where, for any $n \in \mathbb{N}$, $A \subset \mathbb{N}$,

$$A \oplus n \doteqdot \{a + n \mid a \in A\}$$

and

$$A \ominus n \doteqdot \{(a - n) \vee 0 \mid a \in A\}.$$

8.2.2 Proposition *If $\mathcal{G}$ and $\mathcal{F}$ are ultrafilters on $\mathbb{N}$, then*

$$\delta_{\beta\mathbb{N}}(\mathcal{F}, \{\mathcal{G}\}) = n \Leftrightarrow (\mathcal{G} = \mathcal{F} \oplus n \text{ or } \mathcal{G} = \mathcal{F} \ominus n).$$

Proof Fix $\mathcal{F}$ and n. That $\delta_{\beta\mathbb{N}}(\mathcal{F}, \{\mathcal{F} \oplus n\}) \leq n$ follows from the fact that $F \oplus n \subset F^{(n)}$, for all $F \in \mathcal{F}$. To show the other inequality, for any $i \in \{0, 1, \ldots, 2n-1\}$, let

$$Z_i \doteqdot \{2kn + i \mid k \in \mathbb{N}\}.$$

Then there exists a unique $j \in \{0, 1, \ldots, 2n-1\}$ such that $Z_j \in \mathcal{F}$. If we let $j' \doteqdot j + n \pmod{2n}$ then $Z_{j'} \in \mathcal{F} \oplus n$ and

$$\delta_{\beta\mathbb{N}}(\mathcal{F},\{\mathcal{F} \oplus n\}) \geq \inf_{x \in Z_j} \inf_{y \in Z_{j'}} |x - y| = n.$$

This shows that $\delta_{\beta\mathbb{N}}(\mathcal{F},\{\mathcal{F} \oplus n\}) = n$. The result for $\delta_{\beta\mathbb{N}}(\mathcal{F},\{\mathcal{F} \ominus n\})$ is of course an immediate consequence.

To prove the necessity, assume that for some $\mathcal{F}, \mathcal{G} \in \beta\mathbb{N}$ we have $\delta_{\beta\mathbb{N}}(\mathcal{F}, \{\mathcal{G}\}) \leq n < \infty$. Choose $i, j \in \{0, \ldots, 2n\}$ such that

$$\{2n+1\}\mathbb{N} \oplus i \in \mathcal{F}$$

and

$$\{2n+1\}\mathbb{N} \oplus j \in \mathcal{G}.$$

Then

$$\mathcal{B}_\mathcal{F} \doteqdot \{F \cap Z_i \mid F \in \mathcal{F}\}$$

and

$$\mathcal{B}_\mathcal{G} \doteqdot \{G \cap Z_j \mid G \in \mathcal{G}\},$$

are bases for respectively $\mathcal{F}$ and $\mathcal{G}$. For any $A \in \mathcal{B}_\mathcal{F}$, we can find $B \in \mathcal{B}_\mathcal{G}$ such that $B \subset A^{(n)}$. Then there exists a unique $l \in \{-n, \ldots, n\}$ such that $i + l = j \pmod{2n+1}$. It then follows that $B \subset A \oplus l$. Consequently $\mathcal{F} \oplus l \subset \mathcal{G}$ and thus $\mathcal{F} \oplus l = \mathcal{G}$. It now follows from the first part of the theorem that $|l| = n$. $\square$

APPENDIX A

In this appendix we recall basic terminology, concepts, and results in categorical topology. For more detailed information we refer to Adámek *et al.* [1] and Preuss [101].

A.1 BASIC CATEGORICAL CONCEPTS

A.1.1 Definition A *category* $\mathcal{C}$ consists of the following:

1. Objects: a class $|\mathcal{C}|$ of so-called $\mathcal{C}$-*objects*.

2. Morphisms: for each pair (A, B) of $\mathcal{C}$-objects there is a set $\mathcal{C}(A, B)$ (the members of $\mathcal{C}(A, B)$ are called $\mathcal{C}$-*morphisms* from A to B).

3. Identities: for each $A \in |\mathcal{C}|$, there is a $\mathcal{C}$-morphism $1_A \in \mathcal{C}(A, A)$, called the *identity* on A.

4. Composition: for each triple (A, B, C) of $\mathcal{C}$-objects, there is a map $\mathcal{C}(A, B) \times \mathcal{C}(B, C) \longrightarrow \mathcal{C}(A, C) \ : \ (f, g) \longrightarrow g \circ f$, such that the following properties are fulfilled:

(1) if $f \in \mathcal{C}(A, B)$, $g \in \mathcal{C}(B, C)$, and $h \in \mathcal{C}(C, D)$, then $h \circ (g \circ f) = (h \circ g) \circ f$,

(2) for all $B, C \in |\mathcal{C}|$, all $f \in \mathcal{C}(A, B)$, and all $g \in \mathcal{C}(C, A)$: $f \circ 1_A = f$ and $1_A \circ g = g$,

(3) the sets $\mathcal{C}(A, B)$ are pairwise disjoint.

The class of all $\mathcal{C}$-morphisms is denoted by $\mathrm{Mor}\,\mathcal{C}$. If it is clear in which category we are working, then $\mathcal{C}$-morphisms are simply called morphisms. If $f \in \mathcal{C}(A, B)$, we write $f : A \longrightarrow B$ or

$$A \xrightarrow{f} B.$$

A is called the *domain* of f and B is called the *codomain* of f.

A.1.2 Examples 1. The category **TOP** of topological spaces: $|\mathbf{TOP}|$ is the class of all topological spaces and, for all topological spaces X and Y, $\mathbf{TOP}(X,Y)$ is the set of all continuous maps from X to Y. The composition of morphisms is the usual composition of maps.

2. The category **UNIF**: $|\mathbf{UNIF}|$ is the class of all uniform spaces and, for all uniform spaces X and Y, $\mathbf{UNIF}(X,Y)$ is the set of all uniformly continuous maps from X to Y. The composition of morphisms is the usual composition of maps.

3. The category **MET**: $|\mathbf{MET}|$ is the class of all metric spaces and, for all metric spaces X and Y, $\mathbf{MET}(X,Y)$ is the set of all nonexpansive maps from X to Y. The composition of morphisms is the usual composition of maps.

A.1.3 Definition Let $\mathcal{C}$ be a category, $A, B \in |\mathcal{C}|$, and $f \in \mathcal{C}(A,B)$. Then f is called an *isomorphism* provided that there exists a $g \in \mathcal{C}(B,A)$ such that $g \circ f = 1_A$ and $f \circ g = 1_B$. If there exists an isomorphism $f \in \mathcal{C}(A,B)$, then A and B are said to be *isomorphic*.

The morphism g is uniquely determined by f, and is denoted by f^{-1}.

A.1.4 Definition Let $\mathcal{C}$ be a category. A $\mathcal{C}$-morphism $f : A \longrightarrow B$ is called:

1. a *monomorphism* provided that for all pairs (α, β) of $\mathcal{C}$-morphisms with codomain A such that $f \circ \alpha = f \circ \beta$, it follows that $\alpha = \beta$,

2. an *epimorphism* provided that for all pairs (α, β) of $\mathcal{C}$-morphisms with domain B such that $\alpha \circ f = \beta \circ f$, it follows that $\alpha = \beta$,

3. an *extremal monomorphism* provided that f is a monomorphism and, if $f = h \circ g$, where g is an epimorphism, then g is an isomorphism,

4. an *extremal epimorphism* provided that f is an epimorphism and, if $f = g \circ h$, where g is a monomorphism, then g is an isomorphism,

5. a *bimorphism* provided it is at the same time a monomorphism and an epimorphism.

A.1.5 Example The monomorphisms in **TOP** are the injective continuous maps and the epimorphisms are the surjective continuous maps. The extremal monomorphisms in **TOP** are the embeddings and the extremal epimorphisms are the quotient maps.

A.1.6 Definition Let $\mathcal{C}$ be a category and let $A, B \in |\mathcal{C}|$. A is called a *subobject* (respectively *quotient object*) of B provided that $\mathcal{C}(A,B)$ (respectively $\mathcal{C}(B,A)$) contains an extremal monomorphism (respectively extremal epimorphism).

A.1.7 Proposition *Let $\mathcal{C}$ be a category and let $f \in \mathcal{C}(A, B)$. Then the following statements are equivalent:*

1. *f is an isomorphism,*
2. *f is an epimorphism and an extremal monomorphism,*
3. *f is a monomorphism and an extremal epimorphism.*

□

A.1.8 Definition Let $\mathcal{C}$ be a category, I a set, and $(A_i)_{i\in I}$ a family of $\mathcal{C}$-objects. A pair $\big(P, (p_i)_{i\in I}\big)$ with $P \in |\mathcal{C}|$ and $p_i \in \mathcal{C}(P, A_i)$, for each $i \in I$, is called a *product* of the family $(A_i)_{i\in I}$ provided that for each pair $\big(Q, (q_i)_{i\in I}\big)$ with $Q \in |\mathcal{C}|$ and $q_i \in \mathcal{C}(Q, A_i)$, for each $i \in I$, there exists a unique $\mathcal{C}$-morphism $q \in \mathcal{C}(Q, P)$ such that the diagram

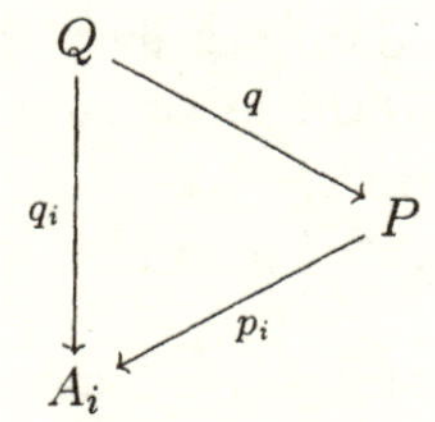

is commutative, for every $i \in I$.

Since products of a family $(A_i)_{i\in I}$ are isomorphic to each other, the object P in the above definition is called *the* product of the family $(A_i)_{i\in I}$ and it is usually denoted by $\prod_{i\in I} A_i$. The morphism p_i is called the i-th projection.

A.1.9 Definition Let $\mathcal{C}$ be a category, I a set, and $(A_i)_{i\in I}$ a family of $\mathcal{C}$-objects. A pair $\big(C, (j_i)_{i\in I}\big)$ with $C \in |\mathcal{C}|$ and $j_i \in \mathcal{C}(A_i, C)$, for each $i \in I$, is called a *coproduct* of the family $(A_i)_{i\in I}$ provided that, for each pair$\big(D, (k_i)_{i\in I}\big)$ with $D \in |\mathcal{C}|$ and $k_i \in \mathcal{C}(A_i, D)$ for each $i \in I$, there exists a unique $\mathcal{C}$-morphism $k \in \mathcal{C}(C, D)$ such that the diagram

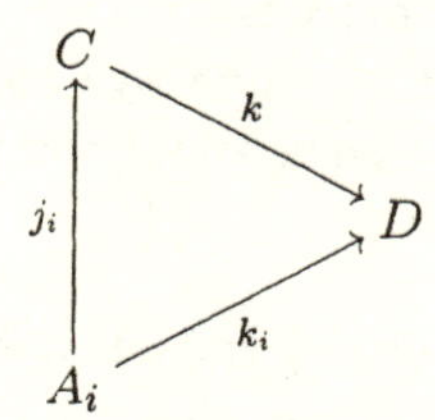

is commutative, for every $i \in I$.

Again, there is only one coproduct of the family $(A_i)_{i\in I}$ up to isomorphism. The object C in the above definition is then also called *the* coproduct of the family $(A_i)_{i\in I}$ and it is usually denoted by $\coprod_{i\in I} A_i$. The morphism j_i is called the i-th injection.

A.1.10 Example The products and coproducts in **TOP** are the topological products and the topological sums, respectively.

A.1.11 Definition Let $\mathcal{C}$ and $\mathcal{D}$ be categories, and let

$$F_1 : |\mathcal{C}| \longrightarrow |\mathcal{D}| \text{ and } F_2 : \mathrm{Mor}\,\mathcal{C} \longrightarrow \mathrm{Mor}\mathcal{D}$$

be maps. Instead of $F_1(A)$ one usually writes $F(A)$, and instead of $F_2(f)$ one usually writes $F(f)$. F is called a *functor* from $\mathcal{C}$ to $\mathcal{D}$ (denoted by $F : \mathcal{C} \to \mathcal{D}$) provided the following conditions are satisfied:

(1) if $f \in \mathcal{C}(A, B)$, then $F(f) \in \mathcal{D}(F(A), F(B))$,

(2) $F(f \circ g) = F(f) \circ F(g)$, whenever $f \circ g$ is defined,

(3) $\forall A \in |\mathcal{C}| : F(1_A) = 1_{F(A)}$.

A.1.12 Definition Let $\mathcal{C}$ and $\mathcal{A}$ be categories. $\mathcal{A}$ is called a *subcategory* of $\mathcal{C}$ if the following conditions are fulfilled:

(1) $|\mathcal{A}| \subset |\mathcal{C}|$,

(2) $\forall A, B \in |\mathcal{A}| : \mathcal{A}(A, B) \subset \mathcal{C}(A, B)$,

(3) the composition of morphisms in $\mathcal{A}$ coincides with the composition of morphisms in $\mathcal{C}$,

(4) for all $\mathcal{A}$-objects A, the identity 1_A is the same in $\mathcal{A}$ and $\mathcal{C}$.

A.1.13 Example Let $\mathcal{C}$ be a category and let $\mathcal{A}$ be a subcategory. Then the inclusion functor $F : \mathcal{A} \to \mathcal{C}$ is defined by $F(A) \doteqdot A$ and $F(f) \doteqdot f$ for every $\mathcal{A}$-object A and every $\mathcal{A}$-morphism f.

A.1.14 Definition Let $F : \mathcal{A} \to \mathcal{B}$ and $G : \mathcal{B} \to \mathcal{C}$ be functors. Then the composition $G \circ F$ of F and G is defined by $(G \circ F)_1 \doteqdot G_1 \circ F_1$ and $(G \circ F)_2 \doteqdot G_2 \circ F_2$. It is obvious that $G \circ F$ is also a functor.

A.1.15 Definition Let $F : \mathcal{C} \to \mathcal{D}$ be a functor.

1. F is called an *embedding* provided that it is injective on morphisms.

2. F is called *faithful* provided that, for all $A, B \in |\mathcal{C}|$, the restrictions $F : \mathcal{C}(A, B) \longrightarrow \mathcal{D}(F(A), F(B))$ are injective.

3. F is called *full* provided that, for all $A, B \in |\mathcal{C}|$, the restrictions $F : \mathcal{C}(A, B) \longrightarrow \mathcal{D}(F(A), F(B))$ are surjective.

4. F is called an *isomorphism* provided that there exists a functor $G : \mathcal{D} \to \mathcal{C}$ such that $G \circ F = I_{\mathcal{C}}$ and $F \circ G = I_{\mathcal{D}}$. If there exists an isomorphism $F : \mathcal{C} \to \mathcal{D}$, then $\mathcal{C}$ is said to be *isomorphic* to $\mathcal{D}$.

A.1.16 Definition Let $\mathcal{C}$ be a category and let $\mathcal{A}$ be a subcategory of $\mathcal{C}$.

1. $\mathcal{A}$ is called a *full subcategory* of $\mathcal{C}$ provided that the inclusion functor $F : \mathcal{A} \to \mathcal{C}$ is full, i.e. $\mathcal{A}(A, B) = \mathcal{C}(A, B)$, for all $\mathcal{A}$-objects A and B.

2. $\mathcal{A}$ is called *isomorphism-closed* provided that every $\mathcal{C}$-object which is isomorphic to an $\mathcal{A}$-object is itself an $\mathcal{A}$-object.

A.1.17 Definition Let $\mathcal{A}$ and $\mathcal{B}$ be categories, $F : \mathcal{A} \to \mathcal{B}$ a functor, and B a $\mathcal{B}$-object.

A pair (u, A) with $A \in |\mathcal{A}|$ and $u : B \longrightarrow F(A)$ a morphism is called a *universal map* for B with respect to F provided that, for each $A' \in |\mathcal{A}|$ and each $f : B \longrightarrow F(A')$, there exists a unique $\mathcal{A}$-morphism $f^* : A \longrightarrow A'$ such that the diagram

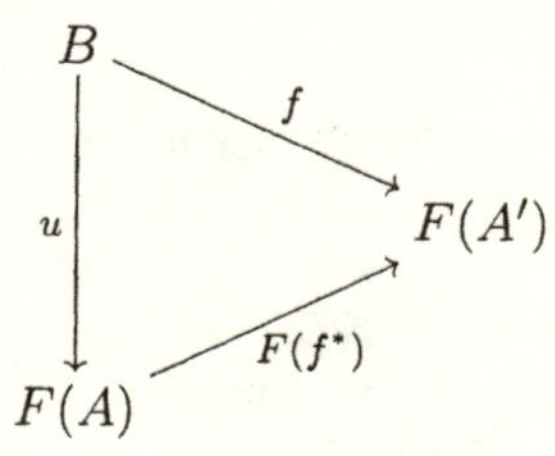

commutes.

A pair (A, u) with $A \in |\mathcal{A}|$ and $u : F(A) \longrightarrow B$ a morphism is called a *co-universal map* for B with respect to F provided that, for each $A' \in |\mathcal{A}|$ and each $f : F(A') \longrightarrow B$, there exists a unique $\mathcal{A}$-morphism $f^* : A' \longrightarrow A$ such

that the diagram

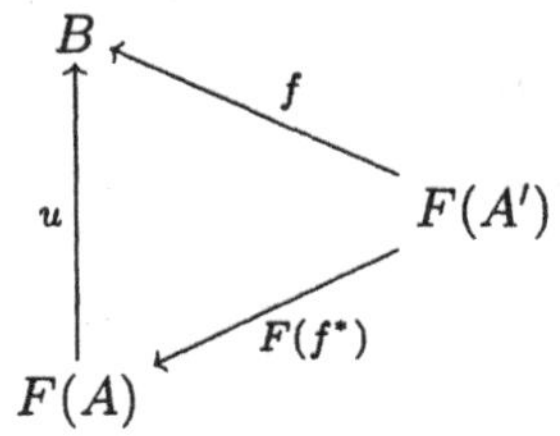

commutes.

A.1.18 Proposition *1. Let* (u, A) *and* (u', A') *be universal maps for a* $\mathcal{B}$*-object* B *with respect to* $F : \mathcal{A} \to \mathcal{B}$*. Then there exists an isomorphism* $f : A \longrightarrow A'$ *such that the diagram*

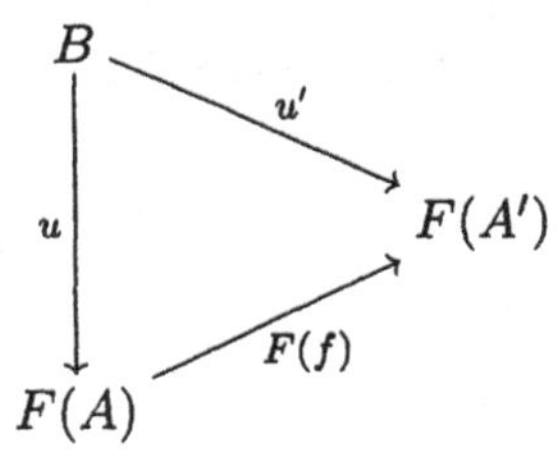

commutes, i.e. a universal map for a $\mathcal{B}$*-object* B *with respect to a functor* $F : \mathcal{A} \to \mathcal{B}$ *is unique up to isomorphism.*

2. Let (A, u) *and* (A', u') *be co-universal maps for a* $\mathcal{B}$*-object* B *with respect to* $F : \mathcal{A} \to \mathcal{B}$*. Then there exists an isomorphism* $f : A \longrightarrow A'$ *such that the diagram*

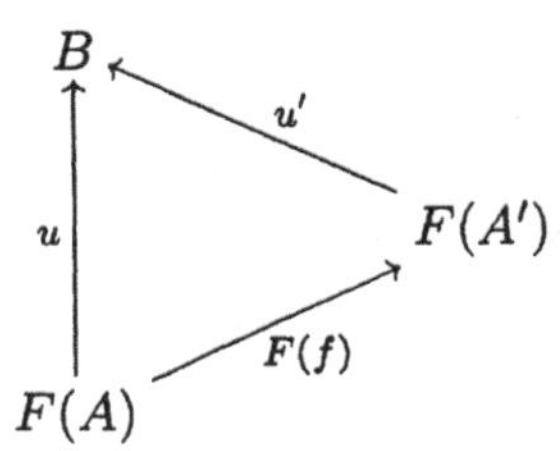

commutes, i.e. a co-universal map for a $\mathcal{B}$*-object* B *with respect to a functor* $F : \mathcal{A} \to \mathcal{B}$ *is unique up to isomorphism.*

□

A.1.19 Definition Let $\mathcal{C}$ and $\mathcal{D}$ be categories and $F : \mathcal{C} \to \mathcal{D}$ and $G : \mathcal{C} \to \mathcal{D}$ be functors.

1. A family $\eta = (\eta_A)_{A \in |\mathcal{C}|}$ such that $\eta_A \in \mathcal{D}(F(A), G(A))$ for every $\mathcal{C}$-object A is called a *natural transformation* (denoted by $\eta : F \to G$) provided that for all $\mathcal{C}$-objects A and B and for each $f \in \mathcal{C}(A, B)$, the diagram

$$\begin{array}{ccc} F(A) & \xrightarrow{\eta_A} & G(A) \\ {\scriptstyle F(f)}\downarrow & & \downarrow{\scriptstyle G(f)} \\ F(B) & \xrightarrow[\eta_B]{} & G(B) \end{array}$$

commutes.

2. A natural transformation $\eta : F \to G$ is called a *natural equivalence* (or a *natural isomorphism*) provided that for each $\mathcal{C}$-object A, η_A is an isomorphism.

3. F and G are called *naturally equivalent* (or *naturally isomorphic*), denoted by $F \approx G$, provided that there exists a natural equivalence from F to G.

A.1.20 Definition Let $F : \mathcal{A} \to \mathcal{B}$ and $G : \mathcal{B} \to \mathcal{A}$ be functors and let

$$v = (v_A)_A : G \circ F \to I_{\mathcal{A}} \text{ and } u = (u_B)_B : I_{\mathcal{B}} \to F \circ G$$

be natural transformations such that:

1. $F(v_A) \circ u_{F(A)} = 1_{F(A)}$ for every $\mathcal{A}$-object A,
2. $v_{G(B)} \circ G(u_B) = 1_{G(B)}$ for every $\mathcal{B}$-object B.

Then G is said to be a *left adjoint* of F, F is said to be a *right adjoint* of G, and (G, F) is called a *pair of adjoint functors*.

A.1.21 Theorem *Let $F : \mathcal{A} \to \mathcal{B}$ be a functor. If, for each $B \in |\mathcal{B}|$, there exists a universal map (u_B, A_B) with respect to F, then there is a unique functor $G : \mathcal{B} \to \mathcal{A}$ such that the following are satisfied:*

1. $G(B) = A_B$ for each $\mathcal{B}$-object B,

2. $u = (u_B)_B : I_{\mathcal{B}} \to F \circ G$ is a natural transformation.

Moreover, there exists a unique natural transformation $v = (v_A)_A : G \circ F \to I_{\mathcal{A}}$ such that the following hold:

(i) $F(v_A) \circ u_{F(A)} = 1_{F(A)}$ for every $\mathcal{A}$-object A,

(ii) $v_{G(B)} \circ G(u_B) = 1_{G(B)}$ for every $\mathcal{B}$-object B,

i.e. G is a left adjoint of F.

If $G : \mathcal{B} \to \mathcal{A}$ is a left adjoint of $F : \mathcal{A} \to \mathcal{B}$ and $u = (u_B)_\mathcal{B} : I_\mathcal{B} \to F \circ G$ is a corresponding natural transformation, then, for each $\mathcal{B}$-object B, $(u_B, G(B))$ is a universal map with respect to F.

□

A.1.22 Proposition *If (G, F) is a pair of adjoint functors, then F preserves products and G preserves coproducts, i.e. if $F : \mathcal{A} \to \mathcal{B}$, then $F\left(\prod_{i\in I} A_i\right) \simeq \prod_{i\in I} F(A_i)$ and $G\left(\coprod_{j\in J} B_j\right) \simeq \coprod_{j\in J} G(B_j)$, for every family $(A_i)_{i\in I}$ of $\mathcal{A}$-objects and every family $(B_j)_{j\in J}$ of $\mathcal{B}$-objects.*

□

A.2 TOPOLOGICAL CONSTRUCTS

A.2.1 Definition A category $\mathcal{C}$ is called a *construct* (or a *concrete category over* **SET**) if there exists a faithful functor $U : \mathcal{C} \to$ **SET**. U is called a *forgetful functor*. For each $\mathcal{C}$-object A, $U(A)$ is called the *underlying set* of A and for each $\mathcal{C}$-morphism f, $U(f)$ is called the *underlying map* of the morphism f.

We can consider a construct $\mathcal{C}$ as a category, with objects structured sets, i.e. pairs (X, ξ) where X is the underlying set of (X, ξ) and ξ is called the $\mathcal{C}$-structure of (X, ξ), and with morphisms suitable maps between X and Y. Composition is the usual composition of maps. We sometimes identify a $\mathcal{C}$-object with its underlying set. This identification is purely notational and it should be clear from the context whether the $\mathcal{C}$-object or its underlying set is meant.

If $\mathcal{C}$ and $\mathcal{D}$ are constructs with forgetful functors $U : \mathcal{C} \to$ **SET** and $V : \mathcal{D} \to$ **SET**, then a functor $F : \mathcal{C} \to \mathcal{D}$ is called *concrete* provided that $V \circ F = U$.

A.2.2 Definition A construct $\mathcal{C}$ is called a *topological construct* if and only if it satisfies the following properties:

(TC1) *Existence of initial structures*: For any set X, any class I, any family $((X_i, \xi_i))_{i\in I}$ of $\mathcal{C}$-objects, and any family $(f_i : X \longrightarrow X_i)_{i\in I}$ of maps, there exists a unique $\mathcal{C}$-structure ξ on X which is *initial* with respect to the source $(f_i : X \longrightarrow (X_i, \xi_i))_{i\in I}$. This means that, for any $\mathcal{C}$-object (Y, η), a map

$g : (Y, \eta) \longrightarrow (X, \xi)$ is a $\mathcal{C}$-morphism if and only if, for every $i \in I$, the map $f_i \circ g : (Y, \eta) \longrightarrow (X_i, \xi_i)$ is a $\mathcal{C}$-morphism.

(TC2) *Fiber-smallness*: For any set X, the $\mathcal{C}$-fiber of X, i.e. the class of all $\mathcal{C}$-structures on X, is a set.

(TC3) *Terminal separator property*: For any singleton X, there exists precisely one $\mathcal{C}$-structure on X.

Let $\mathcal{C}$ be a topological construct. If ξ is the initial structure on X with respect to the source $(f_i : X \longrightarrow (X_i, \xi_i))_{i \in I}$ then $f_i : (X, \xi) \longrightarrow (X_i, \xi_i)$ is a $\mathcal{C}$-morphism, for each $i \in I$.

If X is a set and ξ, η are $\mathcal{C}$-structures on X such that $1_X : (X, \xi) \longrightarrow (X, \eta)$ and $1_X : (X, \eta) \longrightarrow (X, \xi)$ are $\mathcal{C}$-morphisms then $\eta = \xi$.

A.2.3 Definition Let $\mathcal{C}$ be a topological construct, let X be a set, and let ξ and η be $\mathcal{C}$-structures on X. Then ξ is called *finer than* η (and η is called *coarser than* ξ) if and only if $1_X : (X, \xi) \longrightarrow (X, \eta)$ is a $\mathcal{C}$-morphism. This defines a partial order on the $\mathcal{C}$-fiber of X.

A.2.4 Proposition *Let $\mathcal{C}$ be a topological construct. The initial structure ξ on a set X with respect to a source $(f_i : X \longrightarrow (X_i, \xi_i))_{i \in I}$ is the coarsest $\mathcal{C}$-structure on X for which all the maps f_i, $i \in I$, are $\mathcal{C}$-morphisms.*

□

A.2.5 Examples 1. **TOP** and **UNIF** are topological constructs.

2. **MET** is not a topological construct. Initial structures need not exist.

A.2.6 Theorem *In a topological construct $\mathcal{C}$ final structures exist; more precisely, for any set X, any class I, any family $((X_i, \xi_i))_{i \in I}$ of $\mathcal{C}$-objects, and any family $(f_i : X_i \longrightarrow X)_{i \in I}$ of maps, there exists a unique $\mathcal{C}$-structure ξ on X which is final with respect to the sink $(f_i : (X_i, \xi_i) \longrightarrow X)_{i \in I}$. This means that, for any $\mathcal{C}$-object (Y, η), a map $g : (X, \xi) \longrightarrow (Y, \eta)$ is a $\mathcal{C}$-morphism if and only if, for every $i \in I$, the map $g \circ f_i : (X_i, \xi_i) \longrightarrow (Y, \eta)$ is a $\mathcal{C}$-morphism.*

□

Let $\mathcal{C}$ be a topological construct. If $(A_i)_{i \in I}$ is a family of $\mathcal{C}$-objects, indexed by a set I, then the product $\prod_{i \in I} A_i = \left(P, (p_i)_{i \in I}\right)$ exists. Moreover, P is the

Cartesian product of $(A_i)_{i\in I}$, endowed with the initial C-structure with respect to the source $(p_i : P \longrightarrow A_i)_{i\in I}$, where p_i stands for the i-th projection.

Analogously, the coproduct $\coprod_{i\in I} A_i = \left(Q, (j_i)_{i\in I}\right)$ exists. Moreover, Q is the disjoint union of $(A_i)_{i\in I}$, endowed with the final C-structure with respect to the sink $(j_i : A_i \longrightarrow Q)_{i\in I}$, where j_i denotes the i-th injection into the disjoint union.

A.2.7 Proposition *In a topological construct C the final structure on a set X with respect to a sink $(f_i : (X_i, \xi_i) \longrightarrow X)_{i\in I}$ is the finest C-structure for which every map f_i, $i \in I$, is a C-morphism.*

□

A.2.8 Definition Let C be a topological construct and let X be a set. The initial C-structure ξ_i (respectively the final C-structure ξ_d) on X with respect to the empty class is called the *indiscrete structure* (respectively *discrete structure*).

If ξ_i is the indiscrete C-structure on X, then $f : (Y, \eta) \longrightarrow (X, \xi_i)$ is a C-morphism for every C-object (Y, η) and every map $f : Y \longrightarrow X$.

If ξ_d is the discrete C-structure on X, then $f : (X, \xi_d) \longrightarrow (Y, \eta)$ is a C-morphism for every C-object (Y, η) and every map $f : X \longrightarrow Y$.

If (X, ξ) and (Y, η) are objects of a topological construct C and $f : X \longrightarrow Y$ is constant, then $f : (X, \xi) \longrightarrow (Y, \eta)$ is a C-morphism.

A.2.9 Theorem *In a topological construct C a C-morphism $f : (X, \xi) \longrightarrow (Y, \eta)$ is a monomorphism (respectively an epimorphism) if and only if $f : X \longrightarrow Y$ is injective (respectively surjective).*

□

A.2.10 Theorem *Let C be a topological construct and let $f : (X, \xi) \longrightarrow (Y, \eta)$ be a C-morphism.*

1. $f : (X, \xi) \longrightarrow (Y, \eta)$ is an extremal monomorphism if and only if f is an embedding, i.e. $f' : (X, \xi) \longrightarrow \left(f(X), \eta_{f(X)}\right)$ is an isomorphism, where f' is defined by $f'(x) \doteqdot f(x)$ for every $x \in X$ and $\eta_{f(X)}$ is the initial C-structure on $f(X)$ with respect to the inclusion map $i : f(X) \longrightarrow Y$.

2. $f : (X, \xi) \longrightarrow (Y, \eta)$ is an extremal epimorphism if and only if f is a quotient map, i.e. $f : X \longrightarrow Y$ is surjective and η is the final C-structure on Y with respect to $f : (X, \xi) \longrightarrow Y$.

□

A.2.11 Proposition *If* (X, ξ) *is an object of a topological construct* $\mathcal{C}$ *and* $f : X \longrightarrow Y$ *is a bijective map, then there exists a unique* $\mathcal{C}$*-structure* η *on* Y *such that* $f : (X, \xi) \longrightarrow (Y, \eta)$ *is an isomorphism.*

□

A.2.12 Definition A category $\mathcal{C}$ is called *well-powered* provided that, for each $\mathcal{C}$-object X, there is a set $\{m_i : X_i \longrightarrow X \mid i \in I\}$ of monomorphisms which is representative in the following sense: for each monomorphism $m : Y \longrightarrow X$ there is an $i \in I$ and an isomorphism $h_m : Y \longrightarrow X_i$ satisfying $m = m_i \circ h_m$.

A category $\mathcal{C}$ is called *co-well-powered* provided that, for every $\mathcal{C}$-object X, there is a set $\{e_i : X \longrightarrow X_i \mid i \in I\}$ of epimorphisms which is representative in the following sense: for each epimorphism $e : X \longrightarrow Y$ there is an $i \in I$ and an isomorphism $h_e : X_i \longrightarrow Y$ satisfying $e = h_e \circ e_i$.

A.2.13 Theorem *Every topological construct is well-powered and co-well-powered.*

□

A.3 REFLECTIVE AND COREFLECTIVE SUBCONSTRUCTS

We are mainly interested in concrete subcategories of a given construct, i.e. such that the inclusion functor is concrete. Such subcategories are called *subconstructs*.

A.3.1 Definition Let $\mathcal{A}$ be a subcategory of a category $\mathcal{C}$ and $F : \mathcal{A} \to \mathcal{C}$ the inclusion functor.

1. $\mathcal{A}$ is called *reflective* in $\mathcal{C}$ if and only if one of the following equivalent conditions is satisfied:

(i) F has a left adjoint R.

(ii) Each $\mathcal{C}$-object X has a universal map with respect to F. That is, for each $\mathcal{C}$-object X, there exist an $\mathcal{A}$-object $X_{\mathcal{A}}$ and a $\mathcal{C}$-morphism $r_X : X \longrightarrow X_{\mathcal{A}}$ such that for each $\mathcal{A}$-object Y and each $\mathcal{C}$-morphism $f : X \longrightarrow Y$, there is a unique $\mathcal{A}$-morphism $f^* : X_{\mathcal{A}} \longrightarrow Y$ such that $f^* \circ r_X = f$, i.e. such that the following diagram commutes.

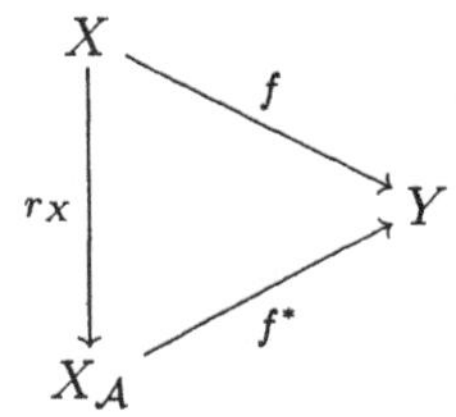

The functor R is called a *reflector*, and $r_X : X \longrightarrow X_{\mathcal{A}}$ is called the *$\mathcal{A}$-reflection* of X.

2. $\mathcal{A}$ is called *epireflective* in $\mathcal{C}$ provided that $\mathcal{A}$ is reflective in $\mathcal{C}$ and, for each $\mathcal{C}$-object X, $r_X : X \longrightarrow X_{\mathcal{A}}$ is an epimorphism. In this case R is called an *epireflector*, and $r_X : X \longrightarrow X_{\mathcal{A}}$ is called the *$\mathcal{A}$-epireflection* of X. Similarly, $\mathcal{A}$ is called *bireflective* in $\mathcal{C}$ provided that $\mathcal{A}$ is reflective in $\mathcal{C}$ and, for each $\mathcal{C}$-object X, $r_X : X \longrightarrow X_{\mathcal{A}}$ is an bimorphism. In this case R is called an *bireflector*, and $r_X : X \longrightarrow X_{\mathcal{A}}$ is called the *$\mathcal{A}$-bireflection* of X.

3. $\mathcal{A}$ is called *coreflective* in $\mathcal{C}$ if and only if one of the following equivalent conditions is satisfied:

(i) F has a right adjoint C.

(ii) Each $\mathcal{C}$-object X has a co-universal map with respect to F. That is, for each $\mathcal{C}$-object X, there exist an $\mathcal{A}$-object $X_{\mathcal{A}}$ and a $\mathcal{C}$-morphism $m_X : X_{\mathcal{A}} \longrightarrow X$ such that, for each $\mathcal{A}$-object Y and each $\mathcal{C}$-morphism $f : Y \longrightarrow X$, there is a unique $\mathcal{A}$-morphism $f^* : Y \longrightarrow X_{\mathcal{A}}$ such that $m_X \circ f^* = f$, i.e. such that the following diagram commutes.

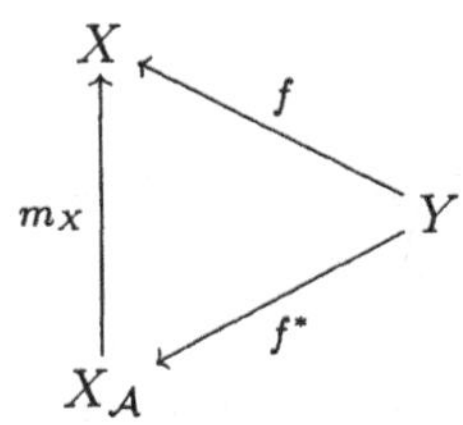

The functor C is called a *coreflector*, and $m_X : X_{\mathcal{A}} \longrightarrow X$ is called the *$\mathcal{A}$-coreflection* of X.

4. $\mathcal{A}$ is called *monocoreflective* in $\mathcal{C}$ if and only if $\mathcal{A}$ is coreflective in $\mathcal{C}$ and, for each $\mathcal{C}$-object X, $m_X : X_{\mathcal{A}} \longrightarrow X$ is a monomorphism. In this case C is called a *monocoreflector*, and $m_X : X_{\mathcal{A}} \longrightarrow X$ is called the *$\mathcal{A}$-monocoreflection* of X. Similarly, $\mathcal{A}$ is called *bicoreflective* in $\mathcal{C}$ if and only if $\mathcal{A}$ is coreflective in

C and, for each C-object X, $m_X : X_A \longrightarrow X$ is a bimorphism. In this case C is called a *bicoreflector*, and $m_X : X_A \longrightarrow X$ is called the *$\mathcal{A}$-bicoreflection* of X.

A.3.2 Theorem *If $\mathcal{A}$ is a full, isomorphism-closed subconstruct of a topological construct C, then the following properties hold:*

1. $\mathcal{A}$ is epireflective in C if and only if $\mathcal{A}$ is closed under the formation of products and subobjects in C,

2. $\mathcal{A}$ is monocoreflective in C if and only if $\mathcal{A}$ is closed under the formation of coproducts and quotient objects in C.

□

A.3.3 Definition An object G is called a *separator* provided that, for each pair of distinct morphisms $f : A \longrightarrow B$ and $g : A \longrightarrow B$, there is a morphism $h : G \longrightarrow A$ such that $f \circ h \neq g \circ h$. An object H is called a *coseparator* provided that, for each pair of distinct morphisms $f : A \longrightarrow B$ and $g : A \longrightarrow B$, there is a morphism $k : B \longrightarrow H$ such that $k \circ f \neq k \circ g$.

Clearly, each object (X, ξ) of a topological construct C such that $X \neq \emptyset$ is a separator and every indiscrete object of a topological construct is a coseparator.

A.3.4 Proposition *1. Let G be a separator of a category C. Then every coreflective, full subconstruct $\mathcal{A}$ of C such that G is an $\mathcal{A}$-object is bicoreflective, i.e. the coreflections are C-bimorphisms.*

2. Let H be a coseparator of a category C. Then every reflective, full subconstruct $\mathcal{A}$ of C such that H is an $\mathcal{A}$-object is bireflective, i.e. the reflections are C-bimorphisms.

□

A.3.5 Corollary *1. A coreflective, full, and isomorphism-closed subconstruct $\mathcal{A}$ of a topological construct C, which contains an object with a nonempty underlying set, is bicoreflective in $\mathcal{A}$.*

2. A reflective, full, and isomorphism-closed subconstruct $\mathcal{A}$ of a topological construct C, which contains an indiscrete C-object, is bireflective in $\mathcal{A}$.

□

A.3.6 Theorem *Every bireflective or bicoreflective, full, and isomorphism-closed subconstruct of a topological construct is a topological construct.*

□

APPENDIX B

B.1 METRIC SPACES

B.1.1 Definition Given a set X, a function

$$d : X \times X \longrightarrow [0, \infty]$$

is called a *metric* if it fulfills the following properties for all $x, y, z \in X$:

1. $d(x, x) = 0$,
2. $d(x, y) \leq d(x, z) + d(z, y)$,
3. $d(x, y) = d(y, x)$,
4. $d(x, y) = 0 \Rightarrow x = y$,
5. $d(x, y) < \infty$.

The second property is referred to as the *triangle inequality*, the third one as *symmetry*, the fourth one as *separatedness*, and the last one as *finiteness*. If d is not necessarily finite, then we call it an *extended metric*, denoted by ∞-metric for short. If it is not necessarily separated, then we call it a *pseudometric*, denoted by p-metric for short, and if it is not necessarily symmetric, then we call it a *quasimetric*, denoted by q-metric for short. Any combination of these is also possible; thus the most general type we consider is an ∞pq-metric, which is a function only fulfilling the first and second properties. The type of function which we consider most often is an ∞p-metric. Note that in Bourbaki [16], what we call an ∞p-metric is called an *écart*, and what we call a metric is called a *distance*. A pair (X, d) where d is an ∞pq-metric on X is called an *extended pseudo-quasimetric space*, ∞pq-metric space for short. Analogous conventions with regard to the use of ∞, p-, and q- hold for spaces. The ball with center x and radius r in the ∞pq-metric space (X, d) is defined and denoted by $B(x, r) \doteqdot \{y \in X \mid d(x, y) < r\}$.

Given two ∞pq-metric spaces, (X, d) and (X', d'), a function $f : X \longrightarrow X'$ is said to be *nonexpansive* if it fulfills the property

$$d'(f(x), f(y)) \leq d(x, y) \qquad \forall x, y \in X.$$

We denote by $pq\mathrm{M}^{\infty}(X)$ (respectively $p\mathrm{M}^{\infty}(X)$) the set of all ∞pq-metrics (respectively all ∞p-metrics) on X. Both these sets, equipped with the pointwise order, are complete lattices, $p\mathrm{M}^{\infty}(X)$ actually being a closed sublattice of $pq\mathrm{M}^{\infty}(X)$.

If $\mathcal{D} \subset pq\mathrm{M}^{\infty}(X)$ (respectively $\mathcal{D} \subset p\mathrm{M}^{\infty}(X)$), then the supremum of $\mathcal{D}$ in $pq\mathrm{M}^{\infty}(X)$ (respectively $p\mathrm{M}^{\infty}(X)$) is the pointwise supremum defined as

$$\left(\sup_{d\in\mathcal{D}} d\right)(x, y) \doteqdot \sup_{d\in\mathcal{D}} d(x, y) \qquad \forall x, y \in X,$$

whereas the infimum of $\mathcal{D}$ in $pq\mathrm{M}^{\infty}(X)$ (respectively $p\mathrm{M}^{\infty}(X)$) is the *metric modification* of the pointwise infimum. If d_0 denotes this pointwise infimum, i.e.

$$d_0(x, y) \doteqdot \inf_{d\in\mathcal{D}} d(x, y) \qquad \forall x, y \in X,$$

then this modification is given by

$$\tilde{d}(x, y) \doteqdot \inf \left\{ \sum_{j=1}^{n} d_0(x_i, x_{i+1}) \mid n \in \mathbb{N}_0,\ x_1 = x,\ x_n = y \right\} \qquad \forall x, y \in X.$$

Thus, $\tilde{d}$ is the largest ∞pq-metric (respectively ∞p-metric) smaller than d_0.

A lattice-theoretical *ideal* in $pq\mathrm{M}^{\infty}(X)$ (respectively $p\mathrm{M}^{\infty}(X)$) is a subset $\mathcal{D}$ which is closed under the formation of taking finite suprema and which is such that if $d \in \mathcal{D}$ and $d' \leq d$ then $d' \in \mathcal{D}$. We call such an ideal an *ideal of ∞pq-metrics* (respectively *ideal of ∞p-metrics*) or an ideal in $pq\mathrm{M}^{\infty}(X)$ (respectively $p\mathrm{M}^{\infty}(X)$).

We call $\mathcal{D}$ *upwards-directed* if for any $d, d' \in \mathcal{D}$ there exists $d'' \in \mathcal{D}$ such that $d \vee d' \leq d''$. If $\mathcal{D}$ is upwards-directed, then obviously, for any finite subset $\mathcal{D}_0 \subset \mathcal{D}$, there exists $d \in \mathcal{D}$ such that $\sup_{e\in\mathcal{D}_0} e \leq d$. An upwards-directed subset of $pq\mathrm{M}^{\infty}(X)$ (respectively $p\mathrm{M}^{\infty}(X)$) is a *basis for an ideal of ∞pq-metrics* (respectively *basis for an ideal of ∞p-metrics*) or basis for an ideal in $pq\mathrm{M}^{\infty}(X)$ (respectively $p\mathrm{M}^{\infty}(X)$) in the lattice-theoretical sense.

A concept which we often encounter and which is somewhat stronger than being upwards directed is the following: if $\mathcal{D}$ is such that, for any finite subset

$\mathcal{D}_0 \subset \mathcal{D}$, we have that $\sup_{d \in \mathcal{D}_0} d \in \mathcal{D}$, then we say that $\mathcal{D}$ is *closed under the formation of finite suprema.*

Clearly an ideal of ∞pq-metrics (respectively ∞p-metrics) is closed under the formation of finite suprema and a collection which is closed under the formation of finite suprema is upwards directed. If $\mathcal{D} \subset pq\mathrm{M}^{\infty}(X)$ (respectively $\mathcal{D} \subset p\mathrm{M}^{\infty}(X)$) is upwards directed, then there exists a smallest ideal of ∞pq-metrics (respectively ∞p-metrics) containing $\mathcal{D}$. This ideal is given by

$$\mathcal{D}' \doteqdot \{d' \in pq\mathrm{M}^{\infty}(X) \mid \exists d \in \mathcal{D} : \ d' \leq d\},$$

respectively

$$\mathcal{D}' \doteqdot \{d' \in p\mathrm{M}^{\infty}(X) \mid \exists d \in \mathcal{D} : \ d' \leq d\}.$$

Extended pseudo-quasimetric spaces and nonexpansive maps form the objects and morphisms of a construct which we denote by $pq\mathbf{MET}^{\infty}$. Obviously the other types of spaces which we have considered give rise to subcategories of $pq\mathbf{MET}^{\infty}$ and here too we adhere to our notational and terminological conventions. The most important subconstruct which we consider is that of ∞p-metric spaces, $p\mathbf{MET}^{\infty}$. Only these two constructs are topological. All other possible subconstructs of $pq\mathbf{MET}^{\infty}$ obtained by combinations of the above properties fail to be topological. If finiteness is assumed, then products and coproducts fail to exist, and if separatedness is assumed, then quotients fail to exist.

B.1.2 Proposition *$pq\mathbf{MET}^{\infty}$ and $p\mathbf{MET}^{\infty}$ are topological constructs.*

□

Initial structures in $pq\mathbf{MET}^{\infty}$ are determined as follows. If X is a set, $(X_j, d_j)_{j\in J}$ is a class of ∞pq-metric spaces, and $f_j : X \longrightarrow X_j$, $j \in J$, is a class of functions, then the ∞pq-metric on X, defined by

$$d : X \times X \longrightarrow [0, \infty] : (x, y) \longrightarrow \sup_{j\in J} d_j(f_j(x), f_j(y)),$$

is initial.

The foregoing result, together with the results from Appendix A, gives rise to the following characterization of products in $pq\mathbf{MET}^{\infty}$. If $(X_j, d_j)_{j\in J}$ is a family of ∞pq-metric spaces, then the product ∞pq-metric d on $\prod_{j\in J} X_j$ is given by

$$d(x, y) = \sup_{j\in J} d_j(\mathrm{pr}_j(x), \mathrm{pr}_j(y)) \qquad \forall x, y \in \prod_{j\in J} X_j.$$

It is clear that the topology underlying an initial ∞pq-metric is, in general, not initial for the topologies underlying the ∞pq-metrics d_j, $j \in J$. The formula for initial structures in $p\mathrm{MET}^\infty$ is precisely the same as the one for $pq\mathrm{MET}^\infty$. Here too, the formation of initial structures, in general, does not commute with the formation of the underlying topologies.

B.1.3 Proposition *$p\mathrm{MET}^\infty$ is a coreflective subconstruct of $pq\mathrm{MET}^\infty$. The $p\mathrm{MET}^\infty$-coreflection of an ∞pq-metric space (X, d) is determined by (X, d^*) where*

$$d^*(x, y) = d(x, y) \vee d(y, x) \qquad \forall x, y \in X.$$

□

B.2 PROBABILISTIC METRIC SPACES

Probabilistic metric spaces were introduced by Menger in [86]. The theory was developed, among others, by Schweizer and Sklar, who wrote a book on the subject in 1983, [106]. The conceptual idea behind probabilistic metric spaces is to provide a mathematical framework for handling imprecise measurements or measurements involving uncertainty. The basic idea here is, rather than to assign real numbers to measurements, to assign probability measures on the positive real line to these measurements. We only recall those concepts which are required in this work. In particular this means we will not introduce probabilistic metric spaces in their full generality as in Schweizer and Sklar [106], but we will restrict our attention to so-called Menger (probabilistic metric) spaces. These are probabilistic metric spaces where the triangle inequality is determined with the aid of a so-called triangular norm or t-norm for short.

B.2.1 Definition A *t-norm* is a function $T : [0, 1] \times [0, 1] \longrightarrow [0, 1]$ fulfilling the following conditions for all $x, y, z \in [0, 1]$:

1. $T(x, y) = T(y, x)$,
2. $T(x, T(y, z)) = T(T(x, y), z)$,
3. $x \leq x', y \leq y' \Rightarrow T(x, y) \leq T(x', y')$,
4. $T(x, 1) = x$.

Typical examples of t-norms are the functions min and T_m where

$$\min(x, y) \doteqdot \min\{x, y\} \text{ and } T_m(x, y) \doteqdot \max\{x + y - 1, 0\} \qquad \forall x, y \in [0, 1].$$

In the context of probabilistic metric spaces and from the technical point of view, rather than using probability measures, it is often better to use distribution functions. By $D([0, \infty])$ we denote the set of all *distribution functions* on $[0, \infty]$, i.e. the set of all functions $F : [0, \infty] \longrightarrow [0, 1]$ fulfilling the following conditions for all $x, y, z \in [0, \infty]$:

1. $x \leq y \;\Rightarrow\; F(x) \leq F(y)$,
2. F is left continuous,
3. $F(0) = 0$,
4. $F(\infty) = 1$.

The particular distribution function defined by

$$[0, \infty] \longrightarrow [0, 1] : x \longrightarrow \begin{cases} 0 & x = 0 \\ 1 & x > 0 \end{cases}$$

is denoted by δ_0.

B.2.2 Definition Given a set X and a t-norm T, a *Menger probabilistic metric (under the t-norm T)* on X is a function

$$\mathfrak{F} : X \times X \longrightarrow D([0, \infty])$$

fulfilling the following conditions for all $x, y, z \in X$ and for all $a, b \in [0, \infty]$:

1. $\mathfrak{F}(x, x) = \delta_0$,
2. $x \neq y \;\Rightarrow\; \mathfrak{F}(x, y) \neq \delta_0$,
3. $\mathfrak{F}(x, y) = \mathfrak{F}(y, x)$,
4. $T(\mathfrak{F}(x, y)(a), \mathfrak{F}(y, z)(b)) \leq \mathfrak{F}(x, z)(a + b)$.

Given a pair of points $x, y \in X$ and a real number $a \in [0, \infty]$, the value $\mathfrak{F}(x, y)(a)$ is interpreted as the probability that the distance between x and y is less than a.

A *Menger probabilistic metric space* is a triple $(X, \mathfrak{F}, T)$ where X is a set, T is a t-norm, and $\mathfrak{F}$ is a Menger probabilistic metric (under the t-norm T).

The *strong uniformity* underlying a probabilistic metric space is the uniformity generated by the entourages

$$\mathbf{U}_{\mathfrak{F}}^{\epsilon} \doteqdot \{(x,y) \in X \times X \mid \mathfrak{F}(x,y)(\epsilon) > 1-\epsilon\},$$

where $\epsilon \in \mathbb{R}_0^+$. The topology generated by this uniformity is called the *strong topology*.

B.3 CONVERGENCE SPACES

The theory of convergence spaces was founded mainly by Choquet [25], Kowalsky [61], Fisher [38] and Kent [60]. We recall the basic definitions.

B.3.1 Definition A *convergence structure* on a set X is a map

$$q : X \longrightarrow 2^{\mathbf{F}(X)}$$

which assigns to every point of X a collection of filters on X and which satisfies the following axioms for all $x \in X$ and for all $\mathcal{F}, \mathcal{G} \in \mathbf{F}(X)$:

1. $\text{stack}x \in q(x)$,
2. $\mathcal{F} \in q(x),\ \mathcal{G} \supset \mathcal{F} \ \Rightarrow\ \mathcal{G} \in q(x)$,
3. $\mathcal{F} \in q(x),\ \mathcal{G} \in q(x) \ \Rightarrow\ \mathcal{F} \cap \mathcal{G} \in q(x)$.

The pair (X,q) is called a *convergence space*. Given convergence spaces (X,q) and (X',q'), a function $f : X \longrightarrow X'$ is called *continuous* if

$$\mathcal{F} \in q(x) \ \Rightarrow \text{stack} f(\mathcal{F}) \in q'(f(x)) \qquad \forall x \in X,\ \forall \mathcal{F} \in \mathbf{F}(X).$$

A convergence structure is called a *pretopology* if, moreover, for any set J and any family of filters $(\mathcal{F}_j)_{j \in J}$ it satisfies

$$(\forall j \in J:\ \mathcal{F}_j \in q(x)) \ \Rightarrow\ \bigcap_{j \in J} \mathcal{F}_j \in q(x).$$

The pair (X,q) is then called a *pretopological space*.

B.3.2 Proposition *If (X, q) is a pretopological space, then*

$$\mathrm{cl}_q : 2^X \longrightarrow 2^X : A \longrightarrow \{x \in X \mid \exists \mathcal{F} \in \mathbf{F}(A) : \ \mathcal{F} \in q(x)\}$$

is a closure operator fulfilling the following properties for all $A, B \in 2^X$:

1. $A \subset \mathrm{cl}_q(A)$,
2. $\mathrm{cl}_q(\emptyset) = \emptyset$,
3. $\mathrm{cl}_q(A \cup B) = \mathrm{cl}_q(A) \cup \mathrm{cl}_q(B)$.

Conversely if $\mathrm{cl} : 2^X \longrightarrow 2^X$ *is a closure operator fulfilling the above three properties, then there exists a unique pretopology q on X such that* $\mathrm{cl} = \mathrm{cl}_q$.

□

Given a filter $\mathcal{F}$ on X and a family of filters $(\mathcal{S}(y))_{y \in X}$ we denote by $\mathcal{D}(\mathcal{S}, \mathcal{F})$ the so-called *diagonal filter* defined as $\bigcup_{F \in \mathcal{F}} \bigcap_{y \in F} \mathcal{S}(y)$. The following proposition characterizes convergence in topological spaces.

B.3.3 Proposition *If (X, q) is a pretopological space such that for any filter $\mathcal{F}$ and any selection of filters $(\mathcal{S}(y))_{y \in X}$*

$$(\mathcal{F} \in q(x), \ \forall y \in X : \ \mathcal{S}(y) \in q(y)) \quad \Rightarrow \quad \mathcal{D}(\mathcal{S}, \mathcal{F}) \in q(x),$$

then q is a topological convergence, i.e. there exists a unique topology $\mathcal{T}$ on X such that $\mathcal{F} \in q(x)$ if and only if $\mathcal{F} \to x$ in $(X, \mathcal{T})$. In terms of the pretopological closure operator, the above condition is equivalent with the idempotency of the closure operator.

□

B.3.4 Proposition *The category* **CONV** *of convergence spaces and continuous maps is a topological construct.*

□

Initial structures in **CONV** are determined as follows. If X is a set, $(X_j, q_j)_{j \in J}$ is a class of convergence spaces, and $f_j : X \longrightarrow X_j$, $j \in J$, is a class of functions, then the convergence structure q on X, defined by

$$\mathcal{F} \in q(x) \text{ if and only if } \forall j \in J : \ \mathrm{stack} f_j(\mathcal{F}) \in q_j(f_j(x)),$$

is initial.

B.3.5 Proposition $\mathbf{PRTOP}$ *is a bireflective subcategory of* $\mathbf{CONV}$. *Given a convergence space* (X, q)*, its* $\mathbf{PRTOP}$*-bireflection is given by*

$$id_X : (X, q) \longrightarrow (X, q^*),$$

where q^* *is the pretopology determined by the closure operator* cl^**, defined as*

$$\mathrm{cl}^*(A) \doteqdot \{x \in X \mid \exists \mathcal{F} \in \mathbf{F}(A) : \ \mathcal{F} \to x\} \qquad \forall A \in 2^X.$$

□

B.3.6 Proposition $\mathbf{TOP}$ *is a bireflective subcategory of* $\mathbf{PRTOP}$. *Given a pretopological space* (X, cl)*, its* $\mathbf{TOP}$*-bireflection is given by*

$$id_X : (X, \mathrm{cl}) \longrightarrow (X, \mathrm{cl}^*),$$

where cl^* *is the closure operator defined by the following transfinite procedure. Given a set* $A \subset X$*, we define*

$$\begin{aligned} &\mathrm{cl}^0(A) \doteqdot A, \\ &\mathrm{cl}^{\alpha+1}(A) \doteqdot \mathrm{cl}(\mathrm{cl}^\alpha(A)) && \alpha \text{ any ordinal}, \\ &\mathrm{cl}^\beta(A) \doteqdot \bigcup_{\alpha<\beta} \mathrm{cl}^\alpha(A) && \beta \text{ a limit ordinal}. \end{aligned}$$

Then $\mathrm{cl}^*(A) \doteqdot \mathrm{cl}^\gamma(A)$ *where* γ *is such that* $\mathrm{cl}^{\gamma+1}(A) = \mathrm{cl}^\gamma(A)$.

□

B.4 QUASI-UNIFORM AND UNIFORM SPACES

Quasi-uniform spaces were originally introduced by Nachbin in 1948 [90] in the context of ordered spaces. As entities in their own right they were introduced and studied by Császár in 1963 [27], who showed that every topological space is quasi-uniformizable. Recently, quasi-uniform spaces have regained interest from researchers working in theoretical computer science, see e.g. Smyth [112].

B.4.1 Definition A *quasi-uniformity* on a set X is a filter $\mathcal{U}$ on $X \times X$ fulfilling the properties:

1. $\forall U \in \mathcal{U} : \ \{(x,x) \mid x \in X\} \subset U$,

2. $\forall U \in \mathcal{U}\ \exists V \in \mathcal{U} : \ V \circ V \subset U$.

If, moreover, the condition

3. $\forall U \in \mathcal{U} : \ U^{-1} \in \mathcal{U}$

is fulfilled, then $\mathcal{U}$ is called a *uniformity*. In the first case the pair $(X,\mathcal{U})$ is called a *quasi-uniform space*, and in the second case it is called a *uniform space*.

Given (quasi-)uniform spaces $(X,\mathcal{U})$ and $(X',\mathcal{U}')$, a function $f : X \longrightarrow X'$ is called *uniformly continuous* if

$$\forall U' \in \mathcal{U}' : \ (f \times f)^{-1}(U') \in \mathcal{U}.$$

B.4.2 Proposition *The category* q**UNIF** *of quasi-uniform spaces and uniformly continuous maps is a topological construct.*

□

Initial structures in q**UNIF** are determined in the same way as in **UNIF**. If X is a set, $(X_j,\mathcal{U}_j)_{j\in J}$ is a class of quasi-uniform spaces, and $f_j : X \longrightarrow X_j$, $j \in J$, is a class of functions, then the quasi-uniformity on X,

$$\mathcal{U} \doteqdot \left\{ V \subset X \times X \mid \exists K \in 2^{(J)}, \forall k \in K\ \exists U_k \in \mathcal{U}_k : \bigcap_{k\in K} (f_k \times f_k)^{-1}(U_k) \subset V \right\},$$

is initial.

B.4.3 Proposition **UNIF** *is a bicoreflective subcategory of* q**UNIF**. *Given a quasi-uniform space* $(X,\mathcal{U})$, *its* **UNIF**-*coreflection is given by*

$$id_X : (X,\mathcal{U}^*) \longrightarrow (X,\mathcal{U})$$

where

$$\mathcal{U}^* \doteqdot \{U \subset X \times X \mid \exists V \in \mathcal{U} : \ V \cap V^{-1} \subset U\}.$$

□

B.5 PROXIMITY SPACES

Proximity spaces were originally introduced by Efremovič [34], [35]. The axioms presented here are as in Naimpally and Warrack [94].

B.5.1 Definition A *proximity* on a set X is a relation $\sim$ on the powerset of X satisfying the following properties for all $A, B \in 2^X$:

1. $A \sim B \Leftrightarrow B \sim A$,
2. $(A \cup B) \sim C \Leftrightarrow A \sim C$ or $B \sim C$,
3. $A \sim B \Rightarrow A \neq \emptyset$ and $B \neq \emptyset$,
4. $A \not\sim B \Rightarrow \exists E \subset X : A \not\sim E$ and $B \not\sim X \setminus E$,
5. $A \cap B \neq \emptyset \Rightarrow A \sim B$.

The pair $(X, \sim)$ is called a *proximity space*.

Given proximity spaces $\left(X, \underset{X}{\sim}\right)$ and $\left(Y, \underset{Y}{\sim}\right)$, a function $f : X \longrightarrow Y$ is said to be a *proximity mapping* if

$$\forall A, B \in 2^X : A \underset{X}{\sim} B \Rightarrow f(A) \underset{Y}{\sim} f(B).$$

B.5.2 Proposition *The category* **PROX** *of proximity spaces and proximity mappings is a topological construct.*

□

Initial structures in **PROX** are determined as follows. If X is a set, $\left(X_j, \underset{j}{\sim}\right)_{j \in J}$ is a class of proximity spaces, and $f_j : X \longrightarrow X_j$, $j \in J$, is a class of functions, then the proximity on X defined by

$$A \sim B$$

if and only if

$$\forall \text{ finite cover } (A_n)_{n \in N} \text{ of } A, \ \forall \text{ finite cover } (B_m)_{m \in M} \text{ of } B$$

$$\exists n \in N, \exists m \in M \text{ such that } \forall j \in J : \ f_j(A_n) \underset{j}{\sim} f_j(B_m)$$

is initial.

BIBLIOGRAPHY

[1] Adámek, J., Herrlich, H., and Strecker, G. E. *Abstract and concrete categories*. John Wiley, New York, (1990).

[2] Anderson, C. L., Hyams, W. H., and McKnight, C. K. Center points of nets. *Can. J. Math. 27* (1975), 418–422.

[3] Arhangel'skii, A., and Wiegandt, R. Connectedness and disconnectedness in topology.
Gen. Topology Appl. 5 (1975), 9–33.

[4] Atsuji, M. Uniform continuity of continuous functions of metric spaces. *Pacific J. Math. 8* (1958), 11–16.

[5] Baekeland, R. *Measures of compactness and their application to convergence of probability measures in approach spaces*. PhD thesis, Vrije Universiteit Brussel, (1993).

[6] Baekeland, R., and Lowen, R. Measures of Lindelöf and separability in approach spaces.
Int. J. Math. Math. Sci. 17 (1994), 597–606.

[7] Baekeland, R., and Lowen, R. Measures of compactness in approach spaces. *Comment. Math. Univ. Carolinae 36* (1995), 327–345.

[8] Banaś, J., and Goebel, K. *Measures of noncompactness in Banach spaces*, vol. 60 of *Lecture Notes in Pure and Applied Mathematics*. Marcel Dekker, New York, (1980).

[9] Beer, G. Convergence of continuous linear functionals and their level sets. *Arch. Math. 52* (1989), 482–491.

[10] Beer, G. Mosco convergence and weak topologies for convex sets and functions.
Mathematika 38 (1991), 89–104.

[11] Beer, G., and Luchetti, R. Weak topologies for the closed subsets of a metrizable space.
Trans. Am. Math. Soc. 335 (1993), 805–822.

[12] Bentley, H. L., Herrlich, H., and Lowen-Colebunders, E. Convergence. *J. Pure Appl. Algebra 68* (1990), 27–45.

[13] Billingsley, P. *Convergence of probability measures.* John Wiley, New York, (1968).

[14] Birkhoff, G. *Lattice theory.* Am. Math. Soc. Colloquium Publications XXV, (1967).

[15] Borges, C. R. On stratifiable spaces. *Pacific J. Math 17* (1966), 1–16.

[16] Bourbaki, N. *Topologie Générale, chapitre 9: Utilisation des nombres réels en topologie générale.* Hermann, Paris, (1958).

[17] Bourbaki, N. *Topologie Générale, chapitres 3 et 4: Groupes topologiques, nombres réels.* Hermann, Paris, (1960).

[18] Bourbaki, N. *Topologie Générale, chapitre 10: Espaces fonctionnels.* Hermann, Paris, (1961).

[19] Brock, P., and Kent, D. C. Approach spaces and probabilistic convergence spaces. *Appl. Categ. Struct..* to appear.

[20] Brock, P., and Kent, D. C. Regularity for approach spaces and probabilistic convergence spaces. *Int. J. Math. Math. Sci..* to appear.

[21] Brümmer, G. C. L. Topological categories. *Topology Appl. 18* (1984), 27–41.

[22] Burke, D. K., and Lutzer, D. J. Recent advances in the theory of generalized metric spaces. In *Topology* (1976), vol. 24 of *Lecture Notes in Pure and Applied Mathematics*, Marcel Dekker, New York, pp. 1–70.

[23] Cantor, G. Über unendliche, lineare Punktmannigfaltigkeiten. *Math. Ann. 21* (1883), 545–591.

[24] Chavez, M. A. Spaces where all continuity is uniform. *Am. Math. Mon. 92* (1985), 487–489.

[25] Choquet, G. Convergences. *Ann. Univ. Grenoble 23* (1947), 57–112.

[26] Costantini, C., Levi, S., and Zieminska, J. Metrics that generate the same hyperspace convergence. preprint.

[27] Császár, A. *Foundations of general topology.* Pergamon Press, Oxford, (1963).

[28] Dieudonné, J. Sur les espaces uniformes complets. *Ann. Ecol. Norm. Supér. 56* (1939), 276–291.

[29] Dikranjan, D., and Tholen, W. *Categorical structure of closure operators.* Kluwer Academic, Dordrecht, (1995).

[30] Dimov, G. D. On κ-metrizable Hausdorff compactifications of κ-metrizable spaces and a new class of spaces, including all separable metrizable spaces. *C R Acad. Bulg. Sci. 36* (1983), 1257–1260.

[31] Dugundji, J. *Topology.* Allyn and Bacon, Boston, (1967).

[32] Edelstein, M. The construction of an asymptotic center with a fixed-point property. *Bull. Am. Math. Soc. 78* (1972), 206–208.

[33] Edelstein, M. Fixed point theorems in uniformly convex Banach spaces. *Proc. Am. Math. Soc. 44* (1974), 369–374.

[34] Efremovič, V. A. Infinitesimal spaces. *Dokl. Akad. Nauk. SSSR 76* (1951), 341–343.

[35] Efremovič, V. A. The geometry of proximity. *Mat. Sb. 31* (1952), 189–200.

[36] Efremovič, V. A., and Švarc, A. S. A new definition of uniform spaces, metrization of proximity spaces. *Dokl. Acad. Nauk. SSSR 89* (1953), 393–396.

[37] Egbert, R. J. Products and quotients in probabilistic metric spaces. *Pacific J. Math. 24* (1968), 437–455.

[38] Fisher, H. Limesraüme. *Math. Ann. 137* (1959), 269–303.

[39] Fletcher, P., and Lindgren, W. F. *Quasi-uniform spaces*, vol. 77 of *Lecture Notes in Pure and Applied Mathematics*. Marcel Dekker, New York, (1982).

[40] Fréchet, M. De l'écart numérique à l'écart abstrait. *Port. Math. 5* (1946), 121–131.

[41] Garkavi, A. L. The best possible net and the best possible cross-section of a set in a normed space. *Am. Math. Soc. Transl. 39* (1964), 111–132.

[42] Gillman, L., and Jerison, M. *Rings of Continuous Functions.* Springer-Verlag, Berlin, (1976).

[43] Gruenhage, G. Generalized metric spaces. In *Handbook of Set-Theoretic Topology*, K. Kunen and J. E. Vaughan, Eds. North-Holland, Amsterdam, (1984), pp. 423–501.

[44] Gruenhage, G. Generalized metric spaces and metrization. In *Recent progress in general topology*, M. Hušek and J. van Mill, Eds. North-Holland, Amsterdam, (1992), pp. 239–274.

[45] Hausdorff, F. *Grundzüge der Mengenlehre*. Veit und Comp., Leipzig, (1914). (reprint, Chelsea, New York, 1949).

[46] Herrlich, H. *Topologische Reflexionen und Coreflexionen*, vol. 78 of *Lecture Notes in Mathematics*. Springer-Verlag, Berlin, (1968).

[47] Herrlich, H. Limit-operators and topological coreflections. *Trans. Am. Math. Soc. 46* (1969), 203–210.

[48] Herrlich, H. On a concept of nearness. *Gen. Topology Appl. 4* (1974), 191–212.

[49] Herrlich, H. *Einführung in die Topologie*. Heldermann-Verlag, Berlin, (1986).

[50] Herrlich, H., and Lowen, R. On simultaneously reflective and coreflective subconstructs. In *G. C. L. Brümmer Festschrift*. to appear.

[51] Herrlich, H., and Strecker, G. E. *Category Theory*. Allyn and Bacon, Boston, (1973). (reprint, Heldermann-Verlag, Berlin, 1979).

[52] Holmes, R. B. *A course in optimization and best approximation*, vol. 257 of *Lecture Notes in Mathematics*. Springer-Verlag, Berlin, (1972).

[53] Horvath, C. Measure of noncompactness and multivalued mappings in complete topological vector spaces. *J. Math. Anal. Appl. 108* (1985), 403–408.

[54] Hušek, M. Reflective and coreflective subcategories of UNIF and TOP. In *Sem. Unif. Spaces* (1973), pp. 113–126.

[55] Isbell, J. R. *Uniform spaces*, vol. 12 of *Math. Surveys*. Am. Math. Soc., (1964).

[56] Isiwata, T. Compact and realcompact κ-metrizable extensions. *Topology Proc. 10* (1985), 95–102.

[57] Isiwata, T. Metrization of additive κ-metric spaces. *Proc. Am. Math. Soc. 100* (1987), 164–168.

[58] Isiwata, T. Continuity of additive κ-metric functions and metrization of κ-metric spaces. *Proc. Am. Math. Soc. 104* (1988), 988–992.

[59] Kannan, V. Reflexive cum coreflexive subcategories in topology. *Math. Ann. 195* (1972), 168–174.

[60] Kent, D. C. Convergence functions and their related topologies. *Fundam. Math. 54* (1964), 125–133.

[61] Kowalsky, J. Limesräume und Komplettierung. *Math. Nachr. 12* (1954), 301–340.

[62] Kuratowski, C. Sur les espaces complets. *Fundam. Math. 15* (1930), 301–309.

[63] Kuratowski, C. *Topology*. Academic Press, (1966). transl.

[64] Lechicki, A., and Levi, S. Wijsman convergence in the hyperspace of a metric space. *Bull. Un. Mat. Ital. 5–B* (1987), 435–452.

[65] Lim, T.-C. On asymptotic centers and fixed points of nonexpansive mappings. *Can. J. Math. 32* (1980), 421–430.

[66] Linz, P. A critique of numerical analysis. *Bull. Am. Math. Soc. 19* (1988), 407–416.

[67] Lowen, E., and Lowen, R. A quasi-topos containing CONV and MET as full subcategories. *Int. J. Math. Math. Sci. 11* (1988), 417–438.

[68] Lowen, E., and Lowen, R. Topological quasitopos hulls of categories containing topological and metric objects. *Cah. Top. Géom. Diff. Catég. 30* (1989), 213–228.

[69] Lowen, E., Lowen, R., and Verbeeck, C. Exponential objects in PRAP. *Cah. Top. Géom. Diff. Catég.*. to appear.

[70] Lowen, R. Kuratowski's measure of noncompactness revisited. *Q. J. Math. Oxford 39* (1988), 235–254.

[71] Lowen, R. Approach spaces: a common supercategory of TOP and MET. *Math. Nachr. 141* (1989), 183–226.

[72] Lowen, R. A topological category suited for approximation theory ? *J. Approximation Theory 56* (1989), 108–117.

[73] Lowen, R. Approximation of weak convergence. *Math. Nachr. 160* (1993), 299–312.

[74] Lowen, R., and Robeys, K. Completions of products of metric spaces. *Q. J. Math. Oxford 43* (1991), 319–338.

[75] Lowen, R., and Robeys, K. Compactifications of products of metric spaces and their relation to Smirnov and Čech-Stone compactifications. *Topology Appl. 55* (1994), 163–183.

[76] Lowen, R., and Sioen, M. Proximal hypertopologies revisited. submitted for publication.

[77] Lowen, R., and Sioen, M. The Wijsman and Attouch-Wets topologies on hyperspaces revisited. *Topology Appl. 70* (1996), 179–197.

[78] Lowen, R., and Vaughan, D. Completion theory for T_0 approach spaces. submitted for publication.

[79] Lowen, R., and Verbeeck, C. Local compactness in approach spaces: part I. *Int. J. Math. Math. Sci.*. to appear.

[80] Lowen, R., and Verbeeck, C. Local compactness in approach spaces: part II. *Int. J. Math. Math. Sci.*. to appear.

[81] Lowen, R., and Windels, B. A-Unif: a common supercategory of UNIF and MET. *Int. J. Math. Math. Sci.*. to appear.

[82] Lowen, R., and Windels, B. On the quantification of uniform properties. submitted for publication.

[83] Lowen, R., and Wuyts, P. A complete classification of the simultaneously bireflective and bicoreflective subconstruct of AP. *Appl. Categ. Struct.*. to appear.

[84] Lowen-Colebunders, E. *Function classes of Cauchy continuous maps*, vol. 123 of *Monographs and Textbooks in Pure and Applied Mathematics*. Marcel Dekker, New York, (1989).

[85] Lowen-Colebunders, E., and Verbeeck, C. Exponential objects in coreflective or quotient reflective subconstructs: a comparison. *Appl. Categ. Struct.*. to appear.

[86] Menger, K. Statistical metrics. *Proc. Natl Acad. Sci. USA 28* (1942), 535–537.

[87] Menger, K. Probabilistic theories of relations. *Proc. Natl Acad. Sci. USA 37* (1951), 178–180.

[88] Michael, E. Topologies on spaces of subsets. *Trans. Am. Math. Soc. 71* (1951), 152–182.

[89] Mrowka, S., and Pervin, W. J. On uniform connectedness.
Proc. Am. Math. Soc. 15 (1964), 446–449.

[90] Nachbin, L. Sur les espaces topologiques ordonnés.
C R Acad. Sci. Paris 226 (1948), 381–382.

[91] Nagata, J. A survey of metrization theory II.
Q & A Gen. Topology 10 (1992), 15–30.

[92] Naimpally, S. A., and Pareek, C. M. Generalized metric spaces via annihilators. Preprint.

[93] Naimpally, S. A., and Pareek, C. M. Characterizations of metric and generalized metric spaces by real valued functions.
Q & A Gen. Topology 8 (1990), 425–439.

[94] Naimpally, S. A., and Warrack, B. D. *Proximity Spaces*.
Cambridge University Press, (1970).

[95] Nishiura, E. Constructive methods in probabilistic metric spaces.
Fundam. Math. 67 (1970), 124–155.

[96] Nussbaum, R. D. A generalization of the Ascoli theorem and an application to functional differential equations.
J. Math. Anal. Appl. 35 (1971), 600–610.

[97] Parthasarathy, K. R. *Probability measures on metric spaces*.
Academic Press, New York, (1967).

[98] Parthasarathy, K. R., and Steerneman, T. A tool in establishing total variation convergence.
Proc. Am. Math. Soc. 94 (1985), 626–630.

[99] Pettis, B. J. Cluster sets of nets.
Proc. Am. Math. Soc. 22 (1969), 386–391.

[100] Preuß, G. E-zusammenhängende Räume.
Manuscr. Math. 3 (1970), 331–342.

[101] Preuß, G. *Theory of Topological Structures*.
Kluwer Academic, Dordrecht, (1987).

[102] Reichel, H. C. Some results on distance functions.
Proceedings of the Fourth Prague Topological Symposium (1976), 371–380.

[103] Reichel, H. C., and Ruppert, W. Über Distanzfunktionen mit Werten in angeordneten Halbgruppen.
Monatsh. Math. 83 (1977), 223–251.

[104] Robeys, K. *Extensions of products of metric spaces*. PhD thesis, Universiteit Antwerpen, RUCA, (1992).

[105] Schweizer, B., and Sklar, A. Statistical metric spaces. *Pacific J. Math. 10* (1960), 313–334.

[106] Schweizer, B., and Sklar, A. *Probabilistic metric spaces*. North-Holland, Amsterdam, (1983).

[107] Shchepin, E. V. On topological products, groups, and a new class of spaces more general than metric spaces. *Sov. Math. Dokl. 17* (1976), 152–155.

[108] Shchepin, E. V. Topology of limit spaces of uncountable inverse spectra. *Russ. Math. Surv. 31* (1976), 155–191.

[109] Shchepin, E. V. On κ-metrizable spaces. *Math. USSR Izv. 14* (1980), 407–440.

[110] Sherwood, H. On E-spaces and their relation to other classes of probabilistic metric spaces. *J. London Math. Soc. 44* (1969), 441–448.

[111] Singer, I. Some relations between dualities, polarities, coupling functionals and conjugations. *J. Math. Anal. Appl. 115* (1986), 1–22.

[112] Smyth, M. B. Quasi-uniformities: reconciling domains with metric spaces. In *Lecture Notes in Computer Science 298* (1987), Springer-Verlag, Berlin.

[113] Suzuki, J., Tamano, K., and Tanaka, Y. κ-metrizable spaces, stratifiable spaces and metrization. *Proc. Am. Math. Soc. 105* (1989), 500–509.

[114] Topsøe, F. *Topology and measure*, vol. 133 of *Lecture Notes in Mathematics*. Springer-Verlag, Berlin, (1970).

[115] Vaughan, J. E. Products of topological spaces. *Gen. Topology Appl. 8* (1976), 207–217.

[116] Vaughan, J. E. Total nets and filters. In *Lecture Notes in Pure and Applied Mathematics*, vol. 24. Marcel Dekker, New York, (1976), pp. 259–265.

[117] Weil, A. Sur les espaces à structure uniforme et sur la topologie générale. *Actual. Sci. Ind. 551* (1937).

INDEX

The manufacturer's authorised representative in the EU for product safety is Oxford University Press España S.A. of el Parque Empresarial San Fernando de Henares, Avenida de Castilla, 2 – 28830 Madrid (www.oup.es/en or product.safety@oup.com). OUP España S.A. also acts as importer into Spain of products made by the manufacturer.

www.ingramcontent.com/pod-product-compliance
Lightning Source LLC
LaVergne TN
LVHW101500200926
848483LV00001B/180
9780198500308